BIOTECHNOLOGY AND THE ENVIRONMENT INCLUDING BIOGEOTECHNOLOGY

BIOTECHNOLOGY AND THE ENVIRONMENT INCLUDING BIOGEOTECHNOLOGY

G. E. ZAIKOV
EDITOR

Nova Science Publishers, Inc.
New York

Production Coordinator: Tatiana Shohov
Coordinating Editor: Tatiana Shohov
Senior Production Editors: Susan Boriotti and Donna Dennis
Production Editors: Marius Andronie and Rusudan Razmadze
Office Manager: Annette Hellinger
Graphics: Levani Chlaidze and Magdalena Nuñez
Editorial Production: Alexandra Columbus, Maya Columbus, Jonathan Boriotti, Robert Brower, Vladimir Klestov, Matthew Kozlowski and Lorna Loperfido
Circulation: Luis Aviles, Raymond Davis, Cathy DeGregory, Melissa Diaz, Ave Maria Gonzalez, Marlene Nuñez, Jeannie Pappas, Vera Popovic and Frankie Punger
Communications and Acquisitions: Serge P. Shohov

Library of Congress Cataloging-in-Publication Data

Biotechnology and the environment including biogeotechnology / G.E. Zaikov, et al., editor.
p. cm.
Includes bibliographical references and index.
ISBN: 1-59454-120-5 (hardcover)
1. Bioremediation.
[DNLM: 1. Biodegradation. 2. Biotechnology. 3. Environment. WA671 B6166 2004]
I. Zaikov, Gennadi Efremovich.

TD192.5625 2004
628.5—dc22

2004014080

400 Oser Ave, Suite 1600
Hauppauge, New York 11788-3619
Tele. 631-231-7269 Fax 631-231-8175
e-mail: Novascience@earthlink.net
Web Site: http://www.novapublishers.com

Printed in the United States of America

CONTENTS

PREFACE

"If you perceive the Proper Way in the morning,
you can quietly die at night
with the sense of followed out duty"
Confucius, China
(500 years A.C.)

"Knowledge is force"
Frances Bacon (1561-1626),
The lord-Chancellor,
Philosopher-materialist
England

We are not sure that anybody sane will willingly agree to die, but everybody wants to know the "Proper Way" in "Science" and someone in "Religion" (and stay alive whenever possible). We also tried to do this on the example of Biotechnology: state of the art and prospects of development.

This volume is proceeding of Second Moscow International Congress on "Biotechnology" which was held on the period of November, 10-14, 2003.

This conference (as well as this volume) included:

- Fundamental researches and biotechnology,
- Biotechnology and medicine,
- Biotechnology and agriculture,
- Biotechnology and industry,
- Biotechnology and environment,
- Biotechnology and food products,
- Biotechnology and biocatalytic technology. Nanotechnology,
- Biogeotechnology and
- Biotechnology and education.

This broad spectrum of fields is very important for Research, Development and Production.

We would like to say many thanks for Dr. Larisa P. Krylova who helped for us very much during preparation of this volume.

The Editor of volume will be happy to receive any kind of scientifical comments about this book. We will use them in our research in future.

Prof. **G.E.Zaikov**
Member of Directorium
N.M.Emanuel Institute of Biochemical Physics
Russian Academy of Sciences.
4 Kosygin st., Moscow- 119991, Russia
Fax 7 095 137 41 01
E-mail chembio@sky,chph.ras.ru

SECTION 1.
BIOTECHNOLOGY

Chapter 1

BIOREMEDIATION OF KAZAKHSTAN SOILS POLLUTED BY 1.1.DIMETHYLHYDRAZINE

***K.K. Boguspaev* [1]*, N.Sh. Alimova* [1]*, S.E. Batyrbekova* [2] *and K.S. Baishev* [2]**
[1]Al-Farabi Kazakh National University, Institute of Biology and Biotechnology
[2]Centre for Physical-Chemical Methods of Analysis
480078, 71, al-Farabi str., Almaty, Kazakhstan

INTRODUCTION

It is known, that (asymmetric) 1.1.dimethylhydrazin (ADMH) in high concentrations renders negative influence on growth and development of plants and human health [1]. Hydrazines, used as high-energy components of rocket fuel, belong to the first class of toxicity. Rocket fuel does regularly pollute the land in Central Kazakhstan with 1.1-dimethylldhydrazines. The concentration of ADMH in soils is very high, more then 10 mg/kg. If marked that ADMH is fixed in soils, water both in plants and their concentrations in various environments is differentiated and depend on combination natural and technogenical factors. In the work it is found that in geochemistry ADMH played an exclusive role in biogeochemical accumulation in soils of humus horizons [2]. ADMH and its derivatives, co-operating by humic acids forms do not represent harm for the environment.

Today there are many data on the negative influence of components of liquid rocket fuel (LRF) on the ecological situation in areas adjoining the spaceport "Baikonur" in Kazakhstan. LRF gets in an environment at the start of a carrier rocket of the "Proton" class, at falling separation steps, at emergencies, and also in storage depots. The amount of the actual fuel and the oxidizer settling on the ground, varies from 0,3 to 2, 0 tons ADMH and from 2 to 4 tons of oxidizing agent (N_2O_4) [3]

The maximum limit concentration (MLC) for air above open sites makes 0.01мг/m^3, for water - 0,02 mg / l, a temporary maximum limit level for soil - 0,1 mg/kg.

Using results of inspections of the spaceport in Kazakhstan, the real concentration of ADMH exceeding maximum concentration limit in soils, plants, ground adjournment, underground and surface waters is determined. Also ADMH is found in animal forages, vegetables, meat of pets that evidence the opportunity of getting into the organism of peoples.

Places for pollution are located and depth of penetration of ADMH on a structure of the ground is determined.

Active research is carried out of the influence ADMH on soils, microflora, and on growth and development of plants with the aim of estimating the degree of oppression of a vegetative cover, irreversibility of processes of degradation of soils and the opportunity of their restoration.

The purpose of work is development of ways for detoxication of ADMH in polluted soil and restoration of biota.

Biohumus

Biohumus produced from residual waste using Californian red worm contains huge amounts of micro- and macromolecules which may act as surfactants, and may replace root exudates. The production of humus is well established in Kazakhstan. In the Kazakh lab, priming of seeds enhanced the speed of emergence and improved field stand and plant growth, both during the vegetative and seed maturation stages, enhancing seedling resistance to salt and osmotic stresses at low temperature. Seed priming plant before sowing may improve not only their growth under stress condition, and also their colonisation with arbuscular visicular mycorrhizal fungi (AM).

Soils Bacteria

Free-living the soil bacteria have ability to fix atmospheric nitrogen. It is known, that an intermediate product of assimilation of atmospheric nitrogen is hydrazine. Under action of special enzyme complex hydrazine of transition to ammonia, which is easily assimilate by plants. Probably, at the first stage of destruction ADMH take part nitrogen-fixing organisms living only in hymus horizons of soils [4].

Plant-Accumulators

Devising a successful phytoremediation strategy necessitates selecting candidate plants with enhanced tolerance to 1.1.dimethilhidrozyne, metals and metal hyperaccumulation (inclusive of uptake, translocation and sequestration).

Representatives of *Amaranthus* species - annual grassy plants with rod, well ramified root system. To the important advantage of Amaranth concern its unpretentiousness to soils, of a high tolerants to draught, salinity, disease and pest. Plasticity and asy adaptation to environment conditions.

In our investigations Amaranth *(A.paniculatus)* on polluted with zinc (Zn) soil, accumulated in vegetative organs up to 1,8 g Zn on 1 kg dry weight of plants [5]. This species of amaranth widely grow in territory of cosmodrome " Baikonur " where ground are strongly polluted with the rests of rocket fuel (1.1.dimethilhidrozyne). Therefore in our experiments species of plants has been used.

Materials and Methods

Definition of Contents ADMG

The experiments used samples of soil polluted with various concentration of 1.1.dimethylhydrazine. Samples selected in places of falling of the first steps of rockets in territory adjoining to spaceport of " Baikonur ".

Maintenance of ADMG in soil made: 100; 500; 1000; and 2000 mg / kg of soil. The characteristic of soil: light-brown loamy, alkalescent soil.

Contents of ADMG determined by a standard technique [4]. In samples of soil (100g) added 100ml of 1N HCl, left for the night for extraction, filtered and distilling at 65-70 ^{0} C. In the distilling solution containing of ADMG added of 1M KOH and 0,6 % of p-NBA (para-nitrobenzaldehyde). For full formation of a complex with ADMG, solution heated to 70-75^{0}C and incubate during 20 minutes. Further solution cooled up to room temperature and extracted by chloroform during 3 minutes at constant mixing. Chloroform deleted on separating funnel and optical density of complex (ADMG-p-NBA) determined at λ = 400nm.

Vegetative Method

As plants - accumulators used *A.paniculatus* , the annual plant growing across all central Kazakhstan [5], and also summer soft wheat (sort Kazakhstan 4) and barley (sort Arna). Samples soil mixed with biohumus in the ratio 50:50 by weight parts. The prepared composition put into the pots (15x13см) on 1 kg. Seeds of an amaranth preliminary wet (priming) in a solution of various concentration of BWH (water extract of biohumus), washed distilled water and dried on air at room temperature. As the control used: a) soil polluted ADMG 500 mg / kg, without biohumus, b). Pure soil (light-brown loam) without ADMG. Phenological examination carried out during 4 months.

Studying of Dynamics the Reduction Concentration of ADMG in Polluted Soil with use of Biohumus

Samples of polluted soil (100g) mixed with various dozes of biohumus (25g, 50g), humidified and incubate at room temperature in darkness. Concentration of ADMG in soil determined at standard method after 2, 24 and 72, 120 hours of incubation.

Results and Discussion

As is known, distribution and accumulation of ADMH in system «soil - plant» depends on a lot of factors: species of plants, structure and types of soils, sources of ADMH , etc. In some cases in general negative action ADMH on soil microflora, plants and animals is denied. At spaceport "Plesetsk (Russia), worked the special commission which has proved absence of any influence ADMH appreciable negatively on surrounding natural environment. In parallel scientists Agrophysic scientific research Institute (Saint-Petersburg, Russia) studied influence ADMH on vegetation of plants and soil microorganisms. During experiments it has been revealed, that ADMH stimulates growth and development of plants, raises their tolerance to stress factors of environment [6]. In a result authors conclude, that plants, together with soil microorganisms, are capable to effectively recycle of ADMH in safe compounds for the people. However, the experiments carried out by us on influence of various concentration of water solutions ADMH on growth and development of an amaranth *(A.paniculatus)*, have shown its oppressing action [7]. In this case we assume, that pure ADMH oppresses development of plants, but in soil ADMH is get mixed up with active components of soil and is neutralized. In result are formed nitrogen-containing components which are used for growth of plants. In our experiments at addition in polluted of ADMH soils of various dozes of biohumus, concentration of ADMH is reduced at the first 24 hours incubation (Table 1).

Table 1.Contents of ADMH in samples of polluted soil at addition of various dozes of biohumus and time of incubation

№ samples of soil	Ratio soil/biohumus	Concentration of ADMH mg / kg		
		24 hours	72 hours	120 hours
663/3	Control	20.6	15.3	15.8
	74 : 25	6.7	6.2	7.0
	50 :50	10.9	6.7	4.7
663/4	Control	10.8	15.3	8.0
	75 : 25	9.1	9.6	6.2
	50 : 50	3.2	3.6	0.7

Apparently from the data of the table at the first 24 hours concentration of ADMH in samples is reduced. Contents of ADMH in soils samples after 24 hours of incubation has decreased on 67.5 %. The next 5 day observed decrease of concentration ADMH, but with less efficacy. In independence of initial concentration of ADMH in samples, after addition of biohumus, there is an appreciable decrease its contents in samples of soil 663/3. Already at addition of 25 % of biohumus there is active neutralization of ADMH both increase in a doze of biohumus and time incubation not strongly influences this process. The same tendency observed at the analysis of samples 663/4. These samples have been selected in various places of range, on a chemical compound of soils were not distinction. For an establishment of the fact bioremediation of polluted soil and removal of residual amounts of ADMH in pots with mixed soil (biohumus and polluted of different concentration of ADMH) have been planted seeds of plant - accumulator TM (heavy metals) *A.paniculatus.*

In figure 1 it is shown, that preliminary wetted in WBH (water solution of biohumus) seeds *A.paniculatus* well germinate on soils with concentration of ADMH up to 100мг/kg, up to 500 mg / kg, up to 1000мг/kg and up to 2000мг/kg.

1 2 3 4 5

Figure 1. Germination of seeds *A.paniculatus* on the soils polluted with various concentration of ADMH with addition of biohumus (50:50.) 1,2 - up to 100 mg / kg. 3-up to 500мг/kg, 4-up to 1000мг,5-up to 2000 mg / kg.

Preliminary wetted BWH seeds, were sowed in pots with the soils polluted with various concentration of ADMH and mixed with dozes (50:50) of biohumus (Figure 1). As is obvious from the figure seeds have germinated in all pots independently from concentration of ADMH. However, after 12 days of growth shoots of amaranth were lost in variant at concentration up to 2000 mg / kg and after 2 months were lost in variant up to 1000мг/kg. (Figure 2).

Further (variant 1 and 2) amaranth plants passed of vegetative phases of development down to formation of reproductive organs. All vegetative period of development of plants passed without noticeable changes in phenotype. (Figure 3).

1 2 3 4

Figure 2. Growth of *A. paniculatus* on polluted soil. 1,2 - concentration of ADMH up to 100мг/kg, 3 - up to 500 mg / kg , 4-up to 2000mg/kg.

1 2 3

Figure 3. Growth of plants *A.paniculatus* on polluted soil.
1,2 - up to 100 mg / kg of ADMH, 3-up to 500 mg / kg ADMH.

Analysis of contents of ADMH in soil carried out after cultivation of plants - accumulators *A.paniculatus,* has shown reduction of its concentration in soil and in 1 and 2 variant concentration of ADMH in soil were of 0.56 mg / kg, in other words contents of ADMH has decreased in more than in 100 times. In variant 4,5 with concentration up to 1000 and 2000mg/kg of ADMH, it is possible to explain the reason death of plants a high level of accumulation of ADMH in vegetative organs of plants. Unfortunately, till now we have no opportunity to determine exact contents of ADMH in vegetative organs of plants and consequently it is necessary to assume indirectly only about ability of an amaranth to accumulate of ADMH in vegetative organs of plants.

Figure 4. The pointer specifies development of reproductive organs

At the following stage of works for bioremediation of soils in the same pots (№ 1.2.3.4.5.6) seeds of wheat sort Kazakhstanskaia 4 and barley sort "Arna", preliminary wetted in WBH, have been planted. Figure 5.

1 2 3 4 5 6

Figure 5. Growth plants of wheat *(T.aestivum)* and barley *(H.vulgare)* on bioremediated soils. 1-control, 2,6, wheat, 3,4,5, - barley

In conclusion it is necessary to note that reliable data was obtained reliable data about reduction of the concentration of ADMH in polluted soil from the territory of the cosmodrome " Baikonur " with addition of various dozes of biohumus.

The obtained results in the future will be used in field experiments for detoxication of ADMH in polluted soils and to restoration of soil biota.

The development of ways of detoxification asymmetric dimetilgidrazin (ADMH) and its derivative by use different concentration of biohumus will be investigated as will biochemical processes for interaction of organic acids contained in biohymus and the effect of nitrogen-fixing organisms on detoxication of ADMH.

The following research will be carried out:

1. Laboratory conditions carrying out the analysis of samples of biohymus with the various contents of hymic acids, on the ability for detoxification of ADMH to establish optimum concentration of biohymus at mixing with samples of soils of the researched region. We will try to determine the presence in soils of samples of nitrogen-fixing organisms and estimate their interaction with ADMH.
2. Realisation of biochemical research of interaction ADMH and its derivative with pure preparations of hymic acids.
3. Bioremediation of soils on experimental fields polluted by ADMH and its derivatives with improvement of biohumus and nitrogen-fixing organisms.

References

A.CH. Avakian: Novel molecular criteria of an estimation of toxic action of derivatives hydrazine. Active forms of oxygen as key agents in the mechanism of toxicity: *Pharmacology and toxicology*. **53**, (1),70-73 (1990).

V.F. Tsvetkov, T.A. Parhimovich, G.N. Djaditcin, T.M. Babich: *Condition of wood vegetation at the cosmodrome "Plesetsk"*. « Environmental problems of the European North », Russian Academy of Science, Ekaterinburg 1996. 290 p. (in Russian).

G.G. Ermakov, G.A. Panova : *Influence of asymmetrical dimethylhydrazin on a condition of soil - vegetative system*. RNCPCH. Saint Petersburg. 1996. 152. (in Russian).

N.S.Kasimov, V.B.Grebenchuk, T.V.Koroleva, J.V.Proskuriakov: Behaviour of components of rocket fuel in soil, waters and plants. *Soil science*, **9**, 110-120. 1994

K.K. Boguspaev, Sh.E. Arystanov, Sh.Sh. Sadykov, A.T.Ivashenko: Accumulation of heavy metals in vegetative organs of amaranth (*A.tricolor, A.paniculatus*) in process of development of plants. *Bulletin KazNU*, **2** (14) 44-53. (2001).

S.A. Sergeev: The report is made at conference "VanPless" - a meeting of experts "Plesetsk" (Russia) and "Vanderberg" (USA), August, 17, 1998. Russian Federation.

K.K. Boguspaev, N.Sh. Alimova, K.S.Baishev, S.B.Batyrbekova:Studying of influence of various concentration of 1.1.dimethylhydrazine (heptyl) on growth and development of plants *Amaranthus paniculatus. Bulletin KazHU, Series ecological*, **2** (11), 31-37.(2002).

Chapter 2

BIO-CALORIMETRIC ANALYSIS OF PHENOL INFLUENCE ON BACTERIA AND THEIR PROTOPLASTS

A.V. Ignatenko, N.A. Belyasova and N.V. Grits
Belorussian State Technological University,
220050, Minsk

ABSTRACT

Phenol action to bacteria and protoplasts of *Bacillus subtilis* and *Clostridium butiricum* was studied by bio-calorimetric techniques. It was shown that bio-calorimetry is well suited to monitoring of bacterial and protoplast's activity, total number of alive cells and is a comfortable indicator of stress, inhibition and toxic effects of xenobiotic.

The problem of discovering of xenobiotics in the environment and foodstuffs making toxic, mutagenic or cancerogenic action is one of the most pressing at present. For regulation of waste movement in the environment it is necessary to have reliable and sensitive methods of dangerous substances testing in any media. The most promising for this matter may be the application of biotests, biosensors and biochips [1] which use different kinds of biological objects as their sensitive elements.

Cell cultures of microorganisms are widely used objects for biotesting of different substances due to of their high sensitivity, simplicity in application, reproduction and economical benefits. A good knowledge of metabolic and genetic processes in many microorganisms makes it possible to apply them for turning up of xenobiotics and studying the mechanisms of their action in live systems.

The main tasks of xenobiotic bio-testing are connected with searching for the most sensitive organisms, estimation some of their biomarker functions and development of the most effective methods and arrangements for their determination.

In the development of new methods of ecological monitoring special attention is given for selection of the most sensitive objects for application. To our mind one of the promising directions of xenobiotic testing may be connected with application of protoplasts as test-objects. They can be used as for direct determination of different substances as for engineering of strains with analytically valuable properties. At present protoplasts are widely used in biotechnology as models for genetic research with bacteria, yeast, plant cells [2]. Protoplasts as test-objects have some advantages before cells. The absence of cell wall assists a penetration of xenobiotics in protoplasts and may increase their sensitivity in comparison with cells. A possibility of cell wall regeneration may be used as one of the bio-marker functions for testing of mutagenic substances.

Among various methods of xenobiotics determination and studying their mechanisms of action at cells one of the most informative is bio-calorimetry. It finds a wide use in biotechnology, ecology, medicine and food industry for express determination of a total number of alive microorganisms, estimation their metabolic activity, sensitivity to any chemicals and studying the mechanisms of cell's respond [3 – 8].

To our knowledge bio-calorimetry hasn't been used for researching of protoplasts. The aim of the present work was checking out the opportunity for bio-calorimetry application for bio-testing of xenobiotics using the example of phenol influence on some bacteria and their protoplasts.

MATERIALS AND METHODS

In this work it was used the strains of bacteria *Bacillus subtilis*, *Clostridium butiricum*, taken from a collection of chair of bio-technology and bio-ecology of the Belorussian State Technological University. The cultures of cells were grown up in full peptone-yeast media (Tetr), containing in 1000 ml of distilled water: 1,5 g of dry nutrient broth, 3 g of yeast extract, 6 g fermentative peptone, pH 7,2. Agarized media was prepared by addition 0,7% or 1,5% agar-agar in Tetr (Tetr-agar) and sterilised 30 min at 0,1 MPa. Protoplasts cultivated in Tetr media with addition of 10 M sucrose (Tetr-osmotic). In experiments were used protoplasts and cells at concentration 10^7 units/ml.

As xenobiotic it was used phenol, purified in accordance with [9]. It was added to the media with protoplasts or cells at the final concentration 0,001-0,01% which was estimated by spectrophotometric method at $\lambda = 270$ nm ($\lg \varepsilon = 3{,}16$) [9].

For receiving of protoplasts it were used bacteria grown in Tetr-media at 30°C during 18 hours. 4 ml of night bacteria cultures was added to 36 ml of fresh media and incubated 2-4 hours at 30 °C with aeration for *Bacillus subtilis and without aeration to Clostridium butiricum*. After that cells were centrifuged during 15 min at 5000 min^{-1} and treated by 1 mg/ml of lysozyme during 30 min at 30°C [2]. The residual enzyme was twice removed by dilution in Tetr-osmotic media and centrifuged 15 min at 5000 min^{-1}. Protoplasts formation efficiency was estimated by plating 10-th dilutions of protoplasts in hypertonic media and in distilled water on the surface of Tetr-agar with and without osmotic stabiliser. The output of protoplasts (P) was estimated as

$$P = [(N_o - N_1)/N_o]\cdot 100\ (\%), \qquad (1)$$

where: N_o - initial total number of cells, N_1 - a number of cells not changed by lysozyme. A total number of cells, protoplasts and cell wall repairing protoplasts were determined by plate account and also by bio-calorimetric method.

Bio-calorimetric analysis were made at microcalorimeter MCM-C (Ukraine). Its preparing and testing were made as described in [8]. The total number of cells or their protoplasts ($N_o(i)$) were estimated as

$$N_o(i) = k(i)\cdot q(i)/F(i), \quad (2)$$

were i - means cells or protoplasts, k - experimental constant, depending from growth of cells during of heat power (q) testing [6], F - heat power per million of cells or protoplasts.

Specific growth rate constant of cells (m) was determined as

$$m = (\ln q_2 - \ln q_1)/\Delta t, \quad (3)$$

were: q_2, q_1 - registered heat power of cells during their growth in time period $\Delta t = t_2 - t_1$.

The total heat production of cells or protoplasts during experiment (Q) was estimated from the experimental power-time curve as

$$Q = \int q(t)\cdot dt \quad (4)$$

Besides using of total heat production (Q), heat power (q) it was also used velocity of heat power change (V_q):

$$V_q = dq/dt. \quad (5)$$

All values were corrected on the shift of base line during their testing.

RESULTS AND DISCUSSION

A heat power-time curve received by microcalorimeter reflects as thermodynamic as kinetic properties of analysed processes [10]. It can be a valuable resource of information about cell's number, their metabolism and mechanisms of xenobiotic action to the cells and their protoplasts.

The amount of heat power produced by cells depends on the number of alive cells, their metabolic activities and can reflect velocity of cells growth. It is possible to appreciate some valuable cell's and protoplast's parameters from their heat power-time curves using equations (2, 3).

We used a bio-calorimetric method for estimation of protoplasts yield and viability in the processes of their receiving. Using of formula (2) it is not difficult to show that from the bio-calorimetric data a yield of protoplasts (P(b)) can be estimated as

$$P(b) = [(q_o - q_1)/q_o]\cdot 100(\%) \quad (6)$$

were q_o, q_1 - heat power of cells before and after protoplasts making.

This value was in a good agreement with P, founded from (1) (r = 0,96). It took only 1 hour to get this value whereas for plate account - some days.

For bio-calorimetric turning up of xenobiotics in the media it was studied phenol influence at bacteria *Clostridium butiricum* and *Bacillus subtilis*. It was used short power-time curves, which reflect initial changes in metabolic processes of cells or their protoplasts induced by phenol. In fig. 1 (a, b) is shown a total heat production of microorganisms after addition of substrate glucose or xenobiotic phenol to cells. Glucose addition stimulated metabolic processes of cells and their heat production has increased. After addition of 0,01% phenol to *Clostridium butiricum* it was also observed activation of heat production (fig.1a, 3), though phenol wasn't their substrate as for *Bacillus subtilis* [11]. Presence of 0,01% phenol in the media turned to decreasing of heat production of *Bacillus subtilis* (fig.1, b) in comparison with control. The total number of alive cells of *Clostridium butiricum* was the same as in control after phenol addition, but in the case of *Bacillus subtilis* it decreased at 20%.

In the table 1 were summarised the results of bio-calorimetric studying of phenol action to the cells of *Clostridium butiricum, Bacillus subtilis* and their protoplasts.

The received bio-calorimetric data show that for protoplasts their thermal parameters heat power and velocity of its change were higher than that for cells.

If to compare a heat production activity of cells and their protoplasts in normal condition it is seen that heat power generated by protoplasts was 1-4 times higher then for cells. It may indicate that they were in stress condition after preparation. As it is known, increasing of heat produced by cells is one of the indicators of their stress conditions [12].

An introduction of xenobiotic in the media led as to increasing as to decreasing of heat power for different micro-organisms, besides that it also changed V_q.

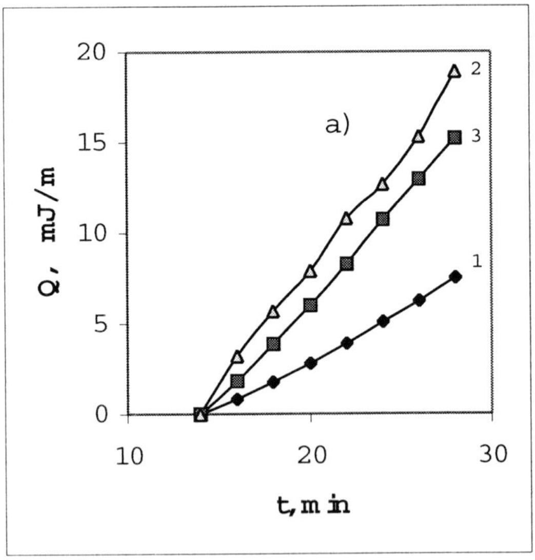

Fig. 1 a. Heat production by microorganisms Clostridium butiricum. Tetr-media. T =30°C 1 - control (without glucose), 2 - after glucose addition (10^{-5} M), 3 - after addition of phenol (0,01%).

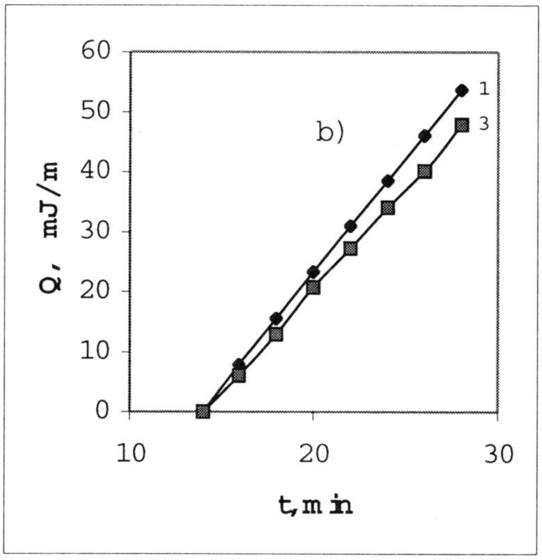

Fig 1 b. Heat production by microorganisms Bacillus subtilis. Tetr-media. T =30°C
1 - control (without glucose), 3 - after addition of phenol (0,01%).

Behaviour of cells and their protoplasts depends on their sensitivity to xenobiotic and the ability to adaptation. As it is known the mechanism of phenol action at the level of cells can be connected with interference of molecules itself or it's quinone products in respiration [13]. It decreases the efficiency of cell's bio-energetic and increases the amount of disappeared heat in environment that can be registered in bio-calorimetric tests.

As it was seen for *Clostridium butiricum* phenol activated metabolic processes of cells and increased their heat production but not killed them. It indicates that these cells were in stress after addition of 0,01% phenol in the media. Analogous activation was also noted for *Bacillus subtilis* at 0,001% concentration of phenol. In spite of phenol is a substrate of *Bacillus subtilis* [11] it decreased their heat production and the number of alive cells at concentration 0,01%. It indicates that possibilities of cells to adaptation were not sufficiently in the case of increasing the amount of xenobiotic in the media.

After addition of phenol to protoplasts their bio-calorimetric parameters heat power and velocity of its change were decreased. The most considerable changes have been seen for protoplasts of *Bacillus subtilis* where V_q had negative value (table 1). Changes of this value correlated with the total amount of alive cells and their conditions. As it is seen from fig. 2 there are some stages of phenol influence at protoplasts of *Bacillus subtilis.* Increasing of V_q value at the first period of time is possible to explain as the stage of stress. Its decreasing (without changes of a total number of alive cells) may be estimated as inhibition. When the V_q value become negative and the total number of alive cells is decreased it means a toxic action of xenobiotic.

As it is seen from table 1 a sensitivity of protoplasts to phenol was higher from that of cells. It connected not only with cell wall absence but due to of changes in their metabolism. A sensitivity of modern microcalorimeters makes it possible to measure 1 μW/ml, that correlates with consumption of 10^{-7} M substrate. It indicates at rather higher sensitivity of bio-calorimetric techniques for xenobiotics detection. Taking into consideration the other profits of bio-calorimetry such as universality for all alive systems, the possibility of analyses of biological objects as in liquid as in solid media it makes bio-calorimetry a valuable technique for xenobiotic testing in environment.

CONCLUSION

Heat production activity is informative function characterizing cells and protoplasts conditions. Bio-calorimetry makes it possible to appreciate a total number of living cells, their physiological activity, presence of xenobiotics in environment. Analyses of short power-time curves gives the opportunity to appreciate a character of xenobiotic action at alive systems (stress condition, inhibitory effect, toxic action) during a short period of time.

Bio-calorimetry is a comfortable method for express estimation of protoplasts yield during their receiving. Heat production activity of protoplasts is higher than cells and they are more sensitive to xenobiotics. It may be connected with their stress condition.

REFERENCES

S.D.Varfolomeev: Biosensors. *Soros educational J.*,**1**, 45 (1997)

K.N.Yakovenko, N.A.Troizkiy: *Protoplasts of microorganisms*. Nayka and Technika, Minsk, 1985.160 p. (in Russian).

L.I. Anatychuk, O.Y. Luste: *Microcalorimetry*. High school, Lvov, 1981. 160p. (in Russian).

V. Hemminger, G. Hene: *Calorimetry: Theory and practice*. Mir, Moscow, 1989. (in Russian).

J. Widdows: Application of calorimetric methods in ecological studies: in A.M. JAMES (ed.): *Thermal and energetic studies of cellular biological systems*. IOP Published Ltd, 182 - 215 (1987).

Standard 27930-88. *Milk and milk products. Biological-calorimetric method for determination of the total bacteria number*. (in Russian).

A.V.Ignatenko: *Microcalorimetric analyses of microorganisms physiological activity in soil and soil extracts: Proceedings of International Scientific Conference "Microbiology and Biotechnology XX1centery"*. Minsk, 22-24 May, 2002, p.222-224. (in Russian).

A.V. Ignatenko, N.V. Griz : Microbiological, organoleptic and visual methods of foodstuffs quality control. *Microcalorimetry. Laboratory manual*. BSTU, Minsk, 2003. 122 p. (in Russian).

A.J.Gordon, R.A.Ford: *The Chemist's Companion. A Handbook of Practical Data, Techniques and References*. -A Wiley-Interscience Publ., N.-Y., L., 1972. 541 p. (in Russian).

A.Synthesis, A.E.Beezer: Thermodynamics and biologically based structure activity relationships: *Thermochimica Acta*, **172**, 81 (1990).

I.M.Gracheva, N.N.Gavrilova, L.A.Ivanova: *Technology of microbial protein substances, aminoacids and fats*. Pish. Prom, Moscow, 1980. 448 p. (in Russian)

I.A.Baskanyan: *Cultivation of microorganisms with mounted properties*. Moscow, 1992. (in Russian).

V.U. Nikonenko et al.: *Chemistry and Technology of water*, **15**, 5 (1993) (in Russian).

Chapter 3

New Strains of Oil Oxidizing Microorganisms Suitable for Purifying Polluted Soils

T.D. Mukasheva, M.Kh. Shigaeva, R. Berzhanova and R. Sydykbekova
The Al-Farabi Kazakh National University
480078 Almaty, Republic of Kazakhstan

Abstract

Eight strains possessing high destructive activity with respect to oil and petroleum products among isolated microorganisms were revealed from oil-cut soils. Morphological, cultural, biochemical and destructive properties of the selected strains were studied. Oil biodestruction in the ground by the strain-destructors was studied in laboratory conditions.

Key words: isolation, oil, petroleum products, polyaromatic hydrocarbons, strain-destructors, hydrophoby, introduction, emulsifying activity, and fluorescence microscopy.

Introduction

Today one of the major environmental problems is pollution of the soil ecosystem by hydrocarbons, first of all by oil and petroleum products. The process of natural autopurification and restoration of oil-cut ecosystems proceeds slowly and measures for acceleration of oil hydrocarbon destruction are necessary [1].

Intensification of biological destruction of hydrocarbons in the environment can be achieved with the help of microorganisms. It is known that purifying of oil-cut soils with the help of microorganisms probably occurs in two ways: activization of natural soil microflora by changes of physical and chemical conditions of the environment and by the introduction of special active oil oxidizing microorganisms into the polluted environment [2].

The isolation of new microbe cultures with high metabolic activity capable of transforming and utilizing oil hydrocarbons and estimation of degradation thus opportunities take on special significance.

MATERIAL AND METHODS

Isolation and Identification of Microorganisms

Microorganisms capable of using oil as a unique source of carbon and energy are isolated from samples of oil-cut soils by method of accumulative cultures. Tengiz oil was used as a substratum. Cultivation was carried out in sample bottles under aeration on a rocking-bar at 24-28^0 C using mineral medium $(NH_4)_2$ SO_4 – 1,0 g, $MgSO_4$ –0, 7 g, NaCl – 0,5 g, K_2HPO_4 – 1,0 g, KH_2PO_4–1,0 g, water - 1 l. Isolation of pure cultures was carried out by cup method on mediums [3]. Morphological, cultural, physiological and biochemical properties according to the modern principles of classification [4, 5, 6] were studied for determination of genus belonging of isolate. Studying of strain-destructors' growth on Tengiz and Kumkol oil and petroleum products was carried out in the liquid mineral medium introducing suspension of three-day cultures. 3, 5, 7, 10, 12 and 15 % oil and petroleum products were added as a unique source of carbon and energy. Sample bottles were put on the rocking-bar (220 rpm) at temperature 25-28 ^{0}C for 10 days. Tengiz oil is light (density - 817,0 kg / m3), sweet crude (sulfur - 0,3 %), small paraffin oil (paraffin - 2,2 %), with insignificant contents of sulfuric acid pitch (4 %), low-viscosity (viscosity is 3,29 mm^2/C at 20^{0}C), asphaltenes - 0,5-3 %, oil chilling temperature is 21^{0}C. Kumkol oil (the density is 8,83 kg / m3) contains 6-10 % of paraffin, 10-16 % of pitches, 52-55 % of pyrobitumen, 20-28 % of oils, oil chilling temperature is 82-85^{0}C. Growth characteristics of the strain-destructors on polyaromatic hydrocarbons were carried out during cultivation on liquid and agarizing mineral mediums. Used concentration of the growth substrata is from 0,5 – 7 %. A filtering paper was impregnated with liquid hydrocarbons: o-xylol, toluol and benzene which then placed on a cover of Petri dish with the medium without source of carbon and energy. For control the same nutrient medium with addition of glucose. Results were taken after 7-10 days of growth.

Methods for Determination Utilised of Oil and Petroleum Products

For determination of oil and petroleum products recycling by a weight method [7] cultures were cultivated in sample bottles with 100 ml of the nutrient medium. Oil and petroleum products were extracted by chloroform with subsequent evaporation on the evaporator, weighing of the residuum and definition of recycling percent. Quantities of the consumed oil and petroleum products were determined by the formula: $A= (a_1-a_2) / a_1 100 \%$ where A - quantity of the consumed oil in %, a_1 - quantity of introduced oil, a_2 - quantity of the residual oil.

Fluorescent Method

Suspension of cells from cultural liquid containing oil and petroleum products as substratum was colored by 0,01 % orange acridine solution according to the standard method

[3]. A luminescent microscope researched the colored specimens with digital camera Leikc DLM.

Definition of Emulsifying Activity

Emulsifying activity of cultural liquids of oil oxidizing microorganisms was determined by the method of Illarionov V.I. and others. [8].

Degree of hydrophoby of the cell surface was determined by the method described by Rosenberg [9].

Coefficient of hydrophobia is calculated by the formula: F=100 % x (Do-D) / Do, where Do - optical density of the cell suspension before hydrocarbon, D - optical density of cell suspensions remained in the water phase after hydrocarbon treatment.

Growth kinetics of microorganisms was studied in periodic culture. Optical density (OD) was determined on PhEC with wavelength 540 nanometers and dish thickness 0,5 cm. Quantity of colony-forming units was determined by the method of serial cultivations with subsequent seeding and calculation of microorganism colonies on the cups with beef-extract agar and wort agar.

Tests of Strain-Destructors in Modeling Experiment

Laboratory works of new strain studying were made on industrially - polluted soil sampled on the territory of Dossor field and containing more than 5 % of oil hydrocarbons. Brown salinity-solonetzic soil of well # 19 was used. The contents of organic carbon is 7,9 %; mobile nitrogen is 25,2 mg/kg; mobile phosphorus is 12,0 mg / kg; mobile potassium is 517,0 mg / kg. Soil is pH 8,7. We took the mixed sample of soil before analyses the soil was cleared from vegetations and bolted through the bolter with holes' diameter 3 mm. Cultures were introduced with mineral salts of nitrogen and phosphorus as K_2HPO_4, NH_4NO_3 was dissolved in water and brought in the soil. Strain – destructors were slurried in water and brought by equal parts in each of exiccators, containing 3 kg of the polluted soil except for control exiccator. Humidity of purified soil had been maintained at 50-60 % of its full moisture capacity for 30 day.

RESULTS AND DISCUSSION

Isolation and Identification of Strain - Destructors

Some variants of the accumulative cultures of the soil samples selected on the territory of Tengiz field were studied in the work. In the first series the accumulative cultures differed in entering the soil into the mineral medium containing as sources of carbon 3, 5, 7 % crude oil. In the second series, oil-oxidizing microorganisms were isolated from the accumulative cultures adapted to high concentrations. For this purpose, 10 ml of inoculum from the accumulative cultures containing 3 % of oil was transferred to the sample bottle containing 5 % of oil. Then there was 10-day incubation and again 10 ml of inoculum was transferred to the sample bottle with 7 % of oil. Pure cultures of the microorganisms capable to grow in the medium with 3%, 5 % and 7 % of Tengiz crude oil were isolated. The maximum quantity of strains capable to grow with 7 %, 10 % and 15 % of oil were isolated from the accumulative cultures adapted to 7 % crude oil. For the further study 7 active isolate were selected. Among these cultures the strains capable to grow in medium with 10 % of Kumkol and with 15 % of Tengiz oil (Table. 1) were found.

Table 1. Growth of strain-destructors on Tengiz and Kumkol oil

Strain	Oil							
	Tengiz				Kumkol			
	Concentration, %							
	10	11	12	15	5	7	10	12
Mycobacterium sp.119-3GM	5	4	3	2	5	5	3	1
Rhodococcus sp. 51 KS	5	4	3	2	5	5	3	1
Bacillus sp. 109 KS	5	4	3	1	5	5	3	1
Pseudomonas sp. 122AS	5	4	3	1	5	5	3	1
Candida sp. V1	4	3	1	1	4	3	1	1
Candida sp. V2	4	3	1	1	4	3	1	1
Trichosporon sp. R20SO2	4	3	1	1	4	4	1	1
Ch2	4	2	1	1	4	3	1	1

Note: growth of cultures is estimated according to the 5-mark system 5 – intensive growth, 4 –good growth, 3 - medium growth, 2 – poor growth, 1- absence of growth.

Genus belonging of strains was determined according to the morphological characteristics, the form and consistence of the colony, pigmentation, and physiological properties. The strain 119-3 ГМ is identified as Mycobacterium thermoresistible, the culture 51 KC is related to the genus of Rhodococcus sp., the strain 109 KC is related to the genus Bacillus sp. and the strain 122 AC is Pseudomonas sp. The yeast cultures B1 and B2 are related to the genus of Candida sp., the strain P20CO2 is determined as Trichosporon sp. The isolate Ч2 is not identified, but according to the sizes and the form of cells it is related to the yeast organisms.

Characteristic of Destructive Activity of Strain-Destructors

Results of the research showed that the most active destructors of Tengiz crude oil in concentration from 3 to 7 % among studied strains were bacterial cultures. They turned out resistant to hydrocarbons of oil and consumption of oil by cultures on the 5th day of growth was from 13,5 ± 2,3 % to 20,2 ± 3,5 %, on the 14th day from 40,64 ± 2,9 % to 58,6 ± 3,5 %, on the 21st day – more 70 % of oil (Figure 1).

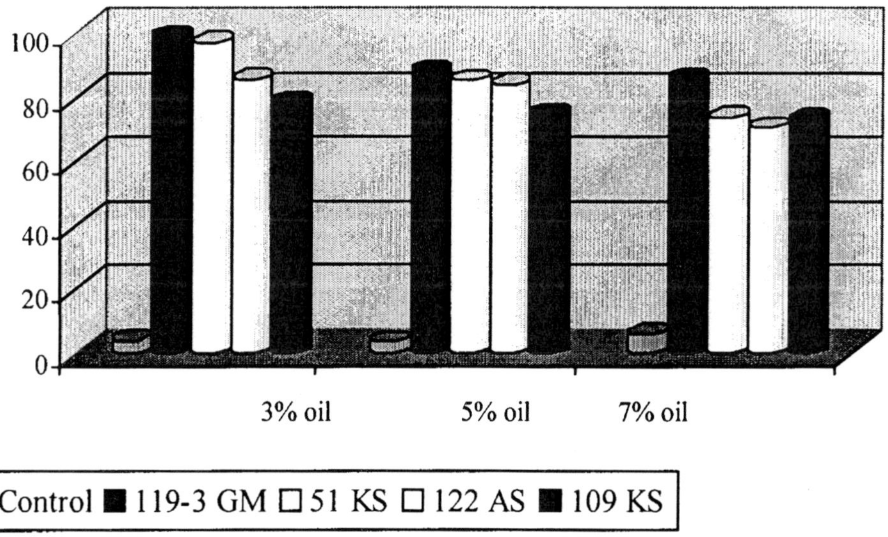

Figure 1. Oil biodestruction by the bacterial and yeast cultures.

Note: Abscissa axis – Tengiz oil, 3%, 5%, 7%, ordinate axis – oil consumption, %.

So, during cultivation on 3 % Tengiz oil the yeast cultures utilize from 52,67 ± 0,14 to 66,28 ± 0,19 % on the 10th day of growth. Consumption of 5 % oil by the cultures was from 33, 25 to 48,35 %, and consumption of 7 % oil by the culture was from 19,27 to 32,12 %.

There was the loss of quantity of oil on the third day during cultivation the active strains on the mediums with Kumkol oil as the unique source of carbon and energy. 3 cultures from the bacterial strains showed the maximum resistance to 5 % Kumkol oil and oil consumption by these cultures on the 5th day was from 12,5 % to 17,2 %, and on the 10th day from 34,8 % to 46,6 %. In comparison with the bacterial cultures the yeast cultures utilize Kumkol oil poorly and they used only from 7,15 to 12,5 % by the 10th day. 3-5 % motor oil, diesel fuel, and gasoline were used during studying the growth ability of the cultures on the mediums with various petroleum products. Basically all cultures showed good ability to oxidize mineral oil (Figure 2).

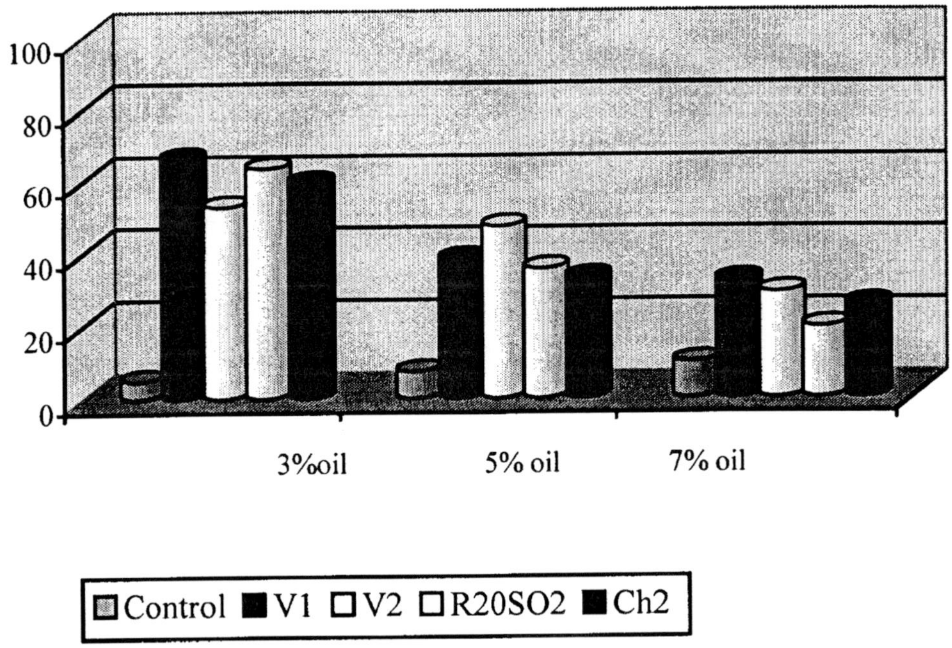

Figure 2. Petroleum product biodestruction by the bacterial cultures

Note: Abscissa axis – 3%, 5%, gasoline, 3%, 5% diesel oil, 3%, 5% motor oil; ordinate axis – utilization of petroleum products, %.

All bacterial strain-destructors were growing intensively on the mediums with concentration 3-5 % of motor oil and diesel fuel. There was the significant gain of biomass on the 5th day and consumption of the motor oil by the cultures was more than 70 %. Significant consumption of all cultures on the 10th day of cultivation was observed on 3 % diesel fuel and was from 41,64 to 96,81 %. The data showed, that recycling of gasoline occurs more full in comparison with diesel fuel and motor oil; it is connected with high natural loss of gasoline. The yeast cultures grew well on the mediums containing 3 % of the petroleum products, recycling changed in limits from 33,35 to 47,21 %. The strain B1 showed maximum activity among the yeast (Figure 3).

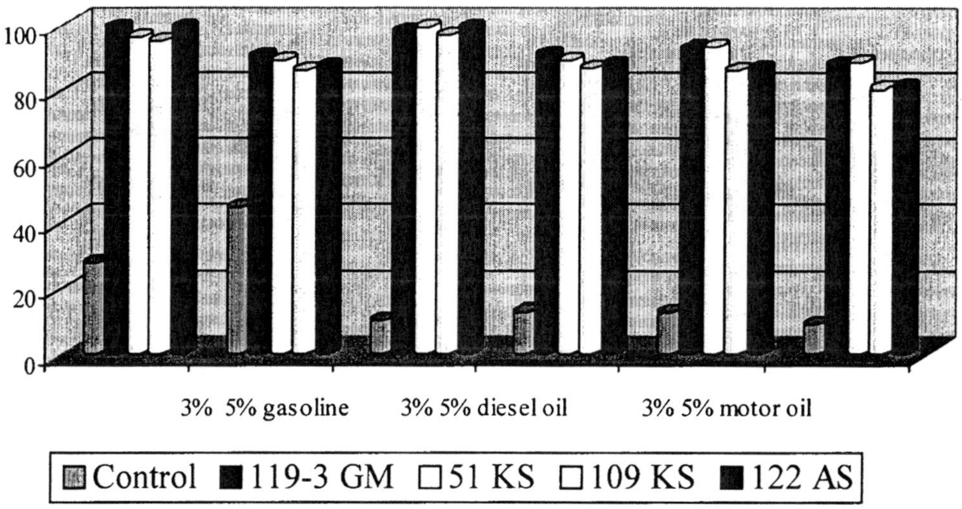

Figure 3. Petroleum product biodestruction be yeast cultures

Note: Abscissa axis – 3%,5% gasoline, 3%,5% diesel oil, 3%,5% motor oil; ordinate axis – utilization of petroleum products, %.

Studying of interaction of microbes - destructors with oil drops and the petroleum products carried out by the fluorescent microscopy method. Specimens of microbe suspensions were studied in dynamics with visual supervision. It was established during the first day, that the separate cells of strain-destructors were adhesive on the drop surface. The activation of process accompanying with increase of number of the adhesive cells on the oil drop surface is visible on the 1st day of the culture evolution. On the 5th day there was active duplication of cells and their penetration into internal part of drops. It is obvious, that the process of penetration promotes extension of the contact surface of the cells with the substratum. Most probably, this process is connected with the hydrophobia of the cell wall [10]. Besides immersing of cells inside drops results in disintegration in fine-dyspersated emulsion. It leads to extend of contact area of the cell of the substratum, and possibly, improves aeration. At the same time increase of intensity of the oil drops' luminescence. The increase of luminescence, maybe, is connected to isolation of bio emulsifier's cells on which dye is adsorbed. On the 7-8th day of cultivation the oil drops turn to the irregular-shaped drops. It is necessary to note that penetration of the cells into the oil drops correlated with the maximal consumption of oil and petroleum products (Figure 4).

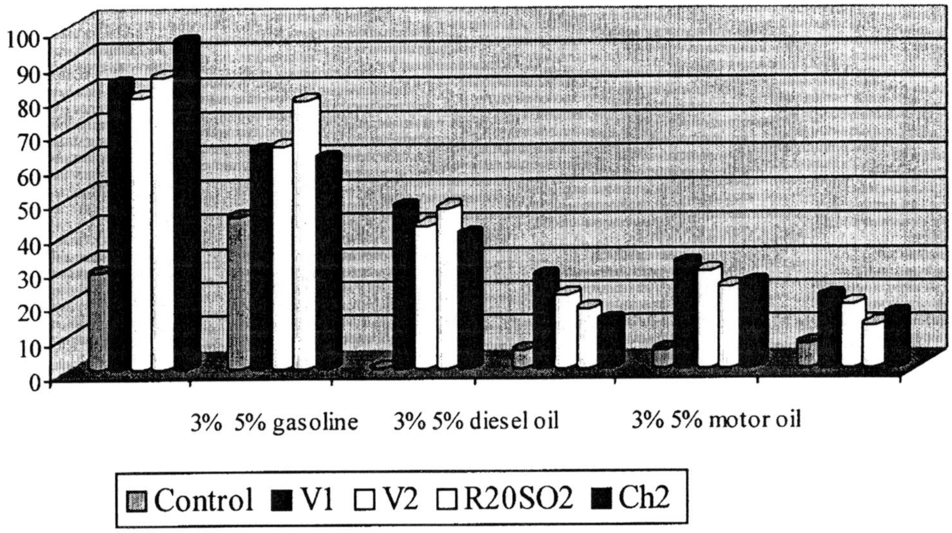

Figure 4. Interaction of bacterial and yeast cells with drops of crude oil

These strains were checked concerning the growth ability on various PAH, their derivatives and metabolites and on aliphatic and monocyclic hydrocarbons (Table. 2). All isolates had high metabolite potential. Except for growth on polyaromatic hydrocarbons: naphthalene, anthracene, phenanthrene, naphthalene 1,5 disulfoacid di-natrium salt, dibenzofuran, diphenyl, paraffin and salicylic acid [11].

Table 2. Growth of the strain-destructors in the medium with various contents of PAH

strain	growth substratum															
	naphthalene,				naphthalene, 1,5				anthracene				phenanthrene			
	concentration, %															
	1	3	5	7	1	3	5	7	1	3	5	7	1	3	5	7
Mycobacterium sp.119-36M	4	3	2	1	5	3	3	1	4	4	3	1	4	4	3	2
Rhodococcussp. 51KS	4	3	2	1	5	3	3	1	4	4	3	1	4	4	3	2
Bacillus sp. 109KS	4	2	1	1	4	2	2	1	4	2	2	1	3	3	2	1
Pseudomonas sp. 122KS	3	2	1	1	4	2	2	1	3	2	2	1	4	3	2	1
Candida sp. V1	4	1	1	1	4	1	1	1	3	1	1	1	4	1	1	1
Candida sp. V2	4	1	1	1	4	1	1	1	3	1	1	1	4	1	1	1
Trichosporon sp. R206O2	5	5	4	1	5	5	4	1	5	5	5	2	5	5	5	2
Ch2	5	5	4	1	4	5	4	1	5	5	5	2	5	5	5	2

Note: growth of cultures is estimated according to the 5-mark system 5 – intensive growth, 4 – good growth, 3 - medium growth, 2 – poor growth, 1- absence of growth.

As studied strain-destructors are isolated from soils with high contents of oil and are isolated from the accumulative cultures containing also high concentration of oil, effect of high concentration of PAH on strain growth was studied. The table 2 shows, that bacterial strains 119-3 ГМ and 51 КС are capable to grow on naphthalene, phenanthrene and anthracene as the unique source of carbon and energy with concentration 1-3 %, growth of these cultures was not observed with higher concentration. 2 strains Р20СО2 and Ч2 of yeast organisms were capable to use phenanthrene, anthracene as growth substratum with concentration 1-5 %. Other cultures used PAH as growth substratum with concentration 0,5-1 % that was typical for the majority of hydrocarbon oxidizing microorganisms [12]

As hydrocarbon oxidizes inside the cell, oil-oxidizing activity of the strains depends on its ability to absorb hydrophobic substratum. The microorganisms capable to use hydrocarbons as the unique source of carbon and energy have special facilities for absorption of these compounds. These facilities include hydrophobia of the cell wall, isolation of bio emulsifiers to the medium and formation of lipophilic channels in the cell wall [12]. Dynamic of hydrophobia parameter's change of the cell surface of strain-destructors conducted on the medium containing 3 % of crude oil. As we can see from (Figure 5) the coefficient is not a constant and varies depending on the cultivation time.

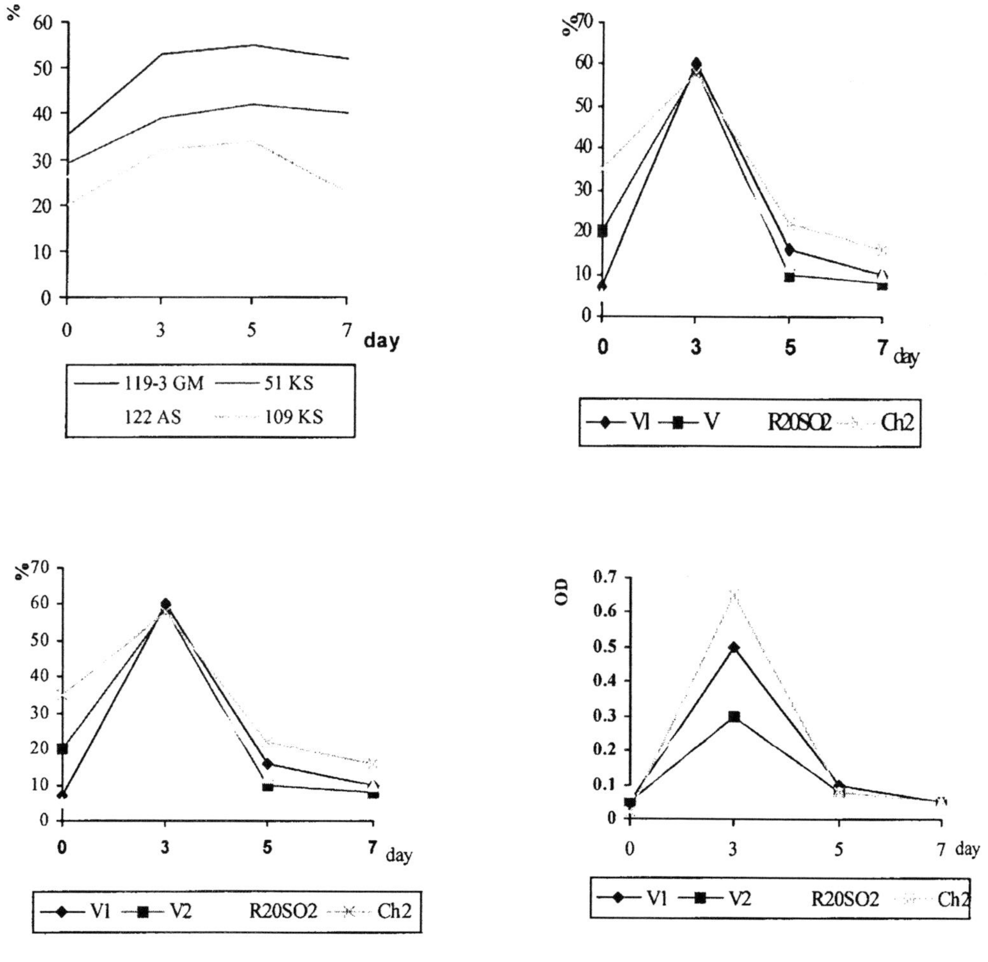

Figure 5. Emulsifying activity (OD) and hydrophobia (%) of the cell wall of the bacterial and yeast culture

Cells of the strain-destructors remained hydrophobic one during all cultivation time. The highest hydrophobia of the cell wall is characteristic for the yeast cultures and it was from 54 to 60 % on the 3rd day of cultivation. Among the bacterial cultures the strain 119-3ГМ showed the high hydrophobia on the 7th day of cultivation. On the 3rd day of cultivation the yeast cultures and on the 5th day of cultivation the bacterial strains showed high emulsifying activity, i.e. there was an emission of PAH into cultural liquid, that is correlated with consumption of oil and petroleum products.

Introduction of Strain-Destructors into the Oil Polluted Soil

Bioremediation by the microorganisms possessing high activity regarding hydrocarbons of oil and capable to utilize is the most perspective method of purifying of oil polluted soils [12]. The results specifying intensification of oil destruction process after introducing mineral power supplies are received at definition of oil volume in the soils. Introducing of the microorganism – destructors with mineral salts into the oil polluted soil, in case of use of 119-3ГМ culture, resulted in biodegradation of 72,31 % of oil on the tenth day and 92,13 % of oil on the thirtieth day. 62,36 and 69,42 % of oil hydrocarbons on the tenth and thirtieth days correspondingly were consumed. In case of introduction the strains 109 KC, 51 KC and 122 AC. 42,71 - 69,5 % and 78,31-87,3 % of oil were subjected to destruction on the tenth and the thirtieth days accordingly in the sample of soil with the solution of mineral salts of nitrogen, phosphorus and yeast cultures. Hence, speed of biological destruction of oil in the polluted soil is highly enough. The maximal decomposition of oil was marked during the first ten days after introduction microorganisms – destructors of the strains 119-3ГМ, P20CO2 and Ч2 (Figure 6).

Thus, as a result of the studies we can assert, that strains with high oil-oxidizing activity were isolated from oil polluted soils. All studied cultures showed resistance to various concentrations of Tengiz and Kumkol oil, to petroleum products and polyaromatic hydrocarbons (naphthalene, anthracene and phenanthrene). Mycobacterium thermoresistible 119-3 ГМ and Rhodococcus 51 KC and the yeast - Trichosporon P20CO2 and Ч2 have high biodestructive activity among the bacterial cultures. So, the culture Mycobacterium thermoresistible 119-3 ГМ utilized more than 70 % of Tengiz oil irrespective of its concentration in the medium.

The biodegradation ability of microorganisms depends on set of factors, including on the mixture of petroleum products, concentration of the petroleum products, growth ability of the microorganisms to adapt in various ecological conditions with high level of oil and petroleum product pollution. Isolated strains Mycobacterium thermoresistible 119-3 ГМ, Trichosporon P20CO2 and Ч2 can use various PAH as the growth substrata and grow with high concentration of naphthalene, anthracene and phenanthrene. According to these data the assumption of an opportunity their participation in degradation of aromatic fractions of oil and petroleum products is made. Also it is possible to assume, that one of the reasons of strains growth on PAH is the synthesis of emulsifiers, biologically active substances. It is determined, that 119-3 ГМ, P20CO2 and Ч2 had not only bio biodegradable, but also high bioemulsifying activity. Besides introduction of strain-destructors 119-3 ГМ, P20CO2 and Ч2 into the oil polluted soil raised a degree of oil destruction and made from the initial contents 96,8 - 84,7 %.

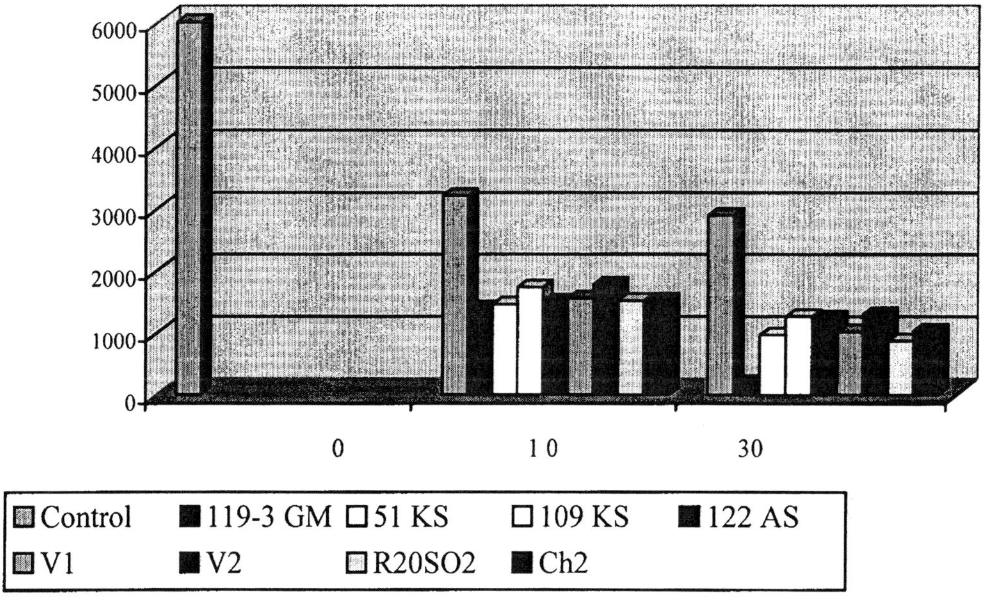

Figure 6. Dynamics of oil biodegradation process after introduction the strain-destructors

Note: abscissa axis – days, ordinate axis - residual contents of oil in the soil, mg/kg

Thus, the studied strain-destructors can be a basis for making the preparations for purifying environmental objects from oil and petroleum products. However further studies are required to find out the conditions which would allow supervising the strain activity at various degrees of impurity.

REFERENCES

1. L.N. Kapotina, G.N.Morschakova. *Biotechnology. Zh.,* **1**. **82** (1998).
2. C. E. Cerniliga *Biodegradation. J.*, **3**, 351. (1992)
3. D.G. Zvyagintsev *Methods of the soil microbiology and biochemistry*. Moscow.1991. p. 13 (in Russian).
4. Williams and Wilkins .Baltimore (Eds.) Bergeys *Manual of Systematic bacteriology* Hong Kong, London, Sydney. 1984-89. v. 2.
5. N.J.W. Kreger – Van Rij *The yeasts. A taxonomic study*. Groningen. The Netherlands. 1984. 1052 p.
6. V.M.Blagodatskaya, L.I.Utkina, I.S.Utkin. "Biology and systematization of concrete groups of microorganisms. Yeast of the genus Candida Berkhout (Systematic, identification)" Pucshino. 1980. p. 290.
7. J.J.Lurje *Analytical chemistry of manufacturing water*. Moscow Chemistry. 1984. p. 448 (in Russian).
8. V.I.Illarionov, N.V.Shishkanova, D.A.Haferburg, T.V.Finogenova *Microbiology. Zh.,* **53**. (3), 423 (1984).

9. M. Rosenberg, D. Gutnick, E. Rosenberg: Adherence of bacteria to hydrocarbons: a simple method for measering cell-surface hydrophobicity // *FEMS J. Microbiol, Lett,* **9**. (1), 29 (1980).
10. N.V.Balashova, I.A.Kosheleva, A.E.Filonov, R.R.Gayazov, A.M.Boronin. *Microbiol. Zh.,* **66** (4), 488 (1997)
11. E.G.Plotnikova, O.V.Altyntseva, I.A.Koshelev, I.F. Puntus, A.E.Filonov, E.J.Gavrish, V.A.Demakov, A.M.Boronin. *Microbiol. Zh.,* **70** (1), 61 (2001).
12. T.V.Koronelli. *Applied Biochemistry and Microbiol. Zh.,* **32** (6), 575 (1996)

Chapter 4

AEROBIC MICROBIAL GRANULE CULTIVATION ON THE BASIS OF *ZOOGLOEA RAMIGERA* AND *SPHAEROTILUS NATANS* PURE CULTURES

G.I. Shaginurova, A.S. Sirotkin, T. Etterer* and P.A. Wilderer*

Kazan State Technological University, 420015 Kazan, ul. K. Marksa ,68,

*Institute of Water Quality Control and Waste Management, 85748 Garching, Germany

ABSTRACT

Aerobic microbial granules were obtained during 2 weeks culturing of flockforming and filament bacteria mixture on a glucose-peptone nutrient medium. The stages of microbial granule formation in a sequencing batch reactor were investigated. It is shown, that only by interaction of these microorganism groops for granules can be formed.

Key Words: Aerobic microbial granules, sequencing batch reactor, Z. ramigera, S. natans

The microbial aggregates known as «microbial granules» represent the spherical biofilm formed in certain conditions by self-immobilisation of activated sludge microorganisms [1]. To increase the oxidizing capacity of biological wastewater treatment plants the use of adhesive and/or adsorptive biomass immobilisation on the surface of inert or active materials is usually recommended. By implementing this method, not only the rise of biomass concentration in a reactor, but also the increase of microorganisms resistance to negative environmental factors, can be achieved [2].

As a rule, special methods of microorganisms agglomeration (chemical binding, floculation etc) are applied by artificial immobilisation when cells themselves are carriers. In this case each cell can be considered as an insoluble basis for other cells, and the concentration of microorganisms in a reactor will be higher than by the immobilisation on the carrier material surface (in the "cell-carrier" system the total weight is divided between cells and the carrier). However, the operational stability of the artificial aggregates depends on physicomechanical characteristics of cell units, but their operational stability parameters are, as a rule, not so good [3]. The formation and development of microbial granules is spontaneous (does not demand the addition of any special agents), in contrast to those of artificial microbial aggregates, and their "mechanical" durability characteristics are comparable with those of biofilms on the carrier.

The formation and development of microbial granules in sewage treatment anaerobic reactors [4, 5] are already investigated and described well enough. The facts of aerobic granules obtaining are still rare: today, only laboratory- or pilot-scale researches using modelling solutions as a microorganisms substratum are carried out. The processes of especially successful microbial granules formation on modelling substrata proceeded in the reactors operated in sequencing batch mode (SBR) with the use of sugar, sodium acetate or ethanol as a source of carbon.

The experiments on sequencing batch cultivation of pure cultures mixtures of structure-forming microorganisms of activated sludge were carried out to find out the mechanism of microbial granules formation.

MATERIALS AND METHODS

As typical structure-forming components of microbial units (activated sludge flocs and microbial granules) filament bacteria *Sphaerotilus natans* (Kutzing 1833, 386AL, ATCC 29329) and flockforming bacteria - *Zoogloea ramigera* (Itzigsohn 1868, synonym *Zoogloea sp.*, ATCC 19544) were used. The bioreactor in volume of 3 dm^3 with an airlift system was used as a basis of the automated laborscale plant (Fig. 1). As a substratum for microorganisms glucose-peptone solution with inorganic salts addition (COD = 470-580 mg/dm^3) was used. During the experiment, the granules were periodically selected from the reactor, and their microbial structure was analyzed by the fluorescent hybridization method [6] and obtaining images by confocal laser microscopy [7].

For the chosen pure cultures, genic probes were used. The genic probes represented complexes, consisting of oligonucleotides, specific to the given kind of microorganisms, and the molecules - markers fluorescing in a certain wave band, namely Sna23A *5'Cy5-CATCCCCCTCTACCGTAC-3'* (the Cy5 marker giving at 633 nm dark blue colouring) and Zora 1414 *5'Cy3-TTCTGGTAAACCCCACTCCC-3 '* (the Cy3 marker giving at 543 nm red colouring). Seedings from pure cultures and from the reactor were periodically made on agar nutrient mediums to check the cultures cleanliness. The hybridization using the wide-spectrum genic probes Euk 516 (to reveal possible infection by fungi or yeast), Eub 338 (to reveal the presence of alien bacteria) was carried out. The absence of the Euk 516 probe signals and the imposing of the Eub 338 probe signals on specific probes signals guaranteed the cleanliness of preparations.

RESULTS AND DISCUSSION

Initial inoculum represented a mixture of large (down to 6-8 mm) *S. natans* flakes and fine (0,5-1 mm) *Z. ramigera* flocs. In the beginning of the cultivation in the airlift reactor, the filaments bacteria dominated. Within the first week of cultivation, the *S. natans* clews preceded the granules (according to the light microscopy (x100) data). During the second week of the experiment, the colonization of clews surface by bacteria *Z. ramigera* was observed (Fig. 2 a, b). The granules, selected from the reactor on the third week, were already dense, smooth spherical formations (1-4 mm) and, as the results of cuts scanning microscopy showed, represented the mixture of two cultures in common matrix (Fig. 3). The obtained granules demonstrated a high consumption efficiency of an organic substratum during a six-hour running cycle - 95-97 % (on COD) and a high speed of sedimentation - down to 1 m/s.

It is necessary to note, that in the pure culture, *Z. ramigera* represent microorganisms, which grow rather slowly, and start to form characteristic flocs, only when the concentration of bacteria is high enough [8]. On the other hand, this kind of bacteria is one of the waste water treatment plants most widespread in activated sludge. The ability of *Z. ramigera* to produce significant amounts of extracellular polymeric substances (EPS) which allows the given kind to carry out flockforming in activated sludge, is well-known. The extracellular polymeric substances, produced by *Z. ramigera,* are represented mainly by polysaccharides. Their structure is stable in a pH-range = 3-10 and by temperatures 25 - 90 ^{0}C (by the temperature decrease, the polymer viscosity raises). As a result, the biopolymers, secreted by Z. ramigera, form viscous, plastic matrix, in which the cells [9] are involved.

By the immobilisation on the *S. natans* flakes, *Z. ramigera* produced extracellular polymeric substances, which promoted the condensation of formed aggregates (the microbial granules). In this case, *S. natans* clews served as carriers, the «carcase» for *Z. ramigera* cells immobilisation. Due to the interaction of both kinds of microorganisms, dense microbial granules were formed. Both kinds of microorganisms are characteristic inhabitants of activated sludge. On the basis of the results obtained, it is possible to make conclusion that these and similar microorganisms is the structure-forming factor for the granules formation from the activated sludge flocs.

Besides, it is known, that the sheaths, forming the *S. natans* filaments, and extracellular polymeric substances, produced by Z. ramigera, have a similar chemical compound [8, 10], due to what they probably form a common matrix, strengthening the structure of a granule.

Granules from activated sludge flocs of municipal wastewater treatment plant were obtained in similar conditions and using similar laboratory system (volume of a reactor of 10 dm^3, except of airlift). After reactor inoculation during the first week of experiment, the condition of active sludge worsened. At a biomass concentration 1,43 g/dm^3 (on dry substance) the sludge volume index increased from 73,5 sm^3/g to 119,2 sm^3/g. The light microscopy observation (x100) showed a rough development of filament bacteria. At this time, the mode with a 1-minute sedimentation stage was applied (Tab. 1). Thus, a slowly settling part of a biomass was washed away with the cleared water, and in the reactor, only fast settling aggregates were selected. In the given conditions, during the second week of experiment, it was already possible to observe in the reactor separate microbial granules (0,5-1 mm) in a mixture with bulking flocs of activated sludge. By the end of the fourth week of experiment, the microbial granules achieved an average diameter 2-4 mm (separate granules achieved a diameter 5-6 mm) and became the prevailing form of microbial units in the reactor. In this period, the organic substratum, ammonium nitrogen and phosphorus consumption data were submitted by microbial granules are represented in Table 2. The next 9 months of work, these parameters varied insignificantly. Within first four months of experiment, the biomass concentration in the reactor (on dry substance) has increased up to 10 g/dm^3. The microbial granules obtained on glucose-peptone substratum did not concede on efficiency of removal of the basic components of modelling wastewater to flocs of activated sludge, but showed the best sedimentation properties (speed of settling about 1 m/s, sludge volume index 20 - 30 sm^3/g).

As a result of the researches carried out, the features of mutual relation of the structure-forming bacterial cultures of activated sludge (*Sphaerotilus natans* and *Zoogloea ramigera)* were found out. An active development of the future microbial granule "carcase" at outstripping consumption by filament bacteria of a carbon substratum with further suppression of their growth by zooglea bacteria, immobilised on the «filament carcase» and involved in slime matrix from EPS, was also detected. The main advantage of microbial granules in comparison with traditional activated sludge flocs in the sequencing batch process of wastewater treatment is their high ability for aggregates sedimentation. At high biomass concentration in the reactor, (when the clearing process efficiency is considerably improved), the granulated biomass subsidence speed is as a rule higher. This results in reducing the time of water treatment common duration.

REFERENCES

J.T.C. Grotenhuis, M. Smit, C.M. Plugge, XU Yuansheng, A.A.M. Van Lammeren, A.J.M. Stams, A.J.B. Zehnder: Bacteriological composition and structure of granular sludge adapted to different substratums. *Appl. Environ. Microbiol.*, **57** (7), 1942-1949 (1991).

A.I. Rodionov, V.N. Klushin, N.S. Torocheshnikov: *Chemistry*, Moscow, 1989. 512 p. (in Russian).

A.P. Sinizyn, E.I. Rainina, V.I. Lozinsky, S.D. Spasov: Moscow State University, Moscow, 1994. 288 p. (in Russian).

G. Lettinga, Pol L.W. Hulshoff: UASB-process design for various types of wastewaters. *Wat. Sci. Tech.*, **24** (8), 87-107 (1991).

S.R Guiot., A Pauss, J.W. Costerton: A structured model of the anaerobic granule consortium. *Wat. Sci. Tech.*, **25** (7), 1-10 (1992).

P. Kaempfer, R. Erhart, C. Beimfohr, M. Wagner, R. Amann: Characterization of bacterial communities from activated sludge: Culture-dependet numerical identification versus in-situ identification using group- and genus-specific rRNA-targeted oligonucleotide probes. *Microb. Ecol.*, (26), 101-121 (1996).

M. Wagner, P. Hutzler, R. Amann: 3-D analysis of complex microbial communities by combining confocal laser scanning microscopy and fluorescence in-situ hybridization (FISH) / In: *Digital image analysis of microbes* (Wilkinson M.H.F., Schut F., eds.): Chichester, England, 467-486 (1998).

J.W. Lee, W.G. Yeomans, A.L. Allen, R.A. Gross, D.L. Kaplan: Production of *zoogloea* gum by *Zoogloea ramigera* with glucose analogs. *Biotech. Lett.*, **19** (8), 799-802 (1997).

K.R. Stauffler, J.G. Leeder, S.S. Wang: Characterization of zooglan 115, an exocellular glycan of *Zoogloea ramigera. J. Food Sci*, (45), 946-952 (1980).

M. Takeda, F. Nakano, T. Nagase, K. Iohara, J.-I. Koizumi: Isolation and chemical composition of the sheath of *Sphaerotilus natans. Biosci. Biotechnol. Biochem.*, **62** (6), 1138-1143 (1998).

Chapter 5

DYNAMICS OF SURFACTANT DESTRUCTION IN COMPLEX PHOTOOZONATION + MICROBIAL DEGRADATION

E.V. Shtamm, N.B. Kozlova, E.V. Alexandrova, S.D. Razumovsky and M.L. Konstantinova
N.M. Emanuel's Institute of Biochemical Physics
Russian Academy of Sciences, Moscow, Russia
J. Emnéus. Sweden, Lund Univ
M. Winther-Nielsen. DHI Water & Environment, Hørsholm, Denmark

In this work surfactant samples of solutions of nonylphenyldodecaetoxylateoxyl (NPE) and sodium salt of decylphenylsulfonic acid (LAS) in distilled water were used. The optimal level for conversion for cleaning processes with biodegradation was established as ~ 25%. Efficiency of detoxication of wastewaters samples was controlled by biotesting methods: by survival of crustaceous (*Ceriodaphnia dubia*) (1) and by inhibition of chemotaxic reaction of infusoria (*Tetrahymena pyriformis*) (2). The values LC50 were 5mg/L for LAS and 13 mg/L for NPE (1), IC50 were 2,5mg/L for LAS and 8,5mg/L for NPE (2).

Keywords: surfactants, photoozonation, biodegradation, biotesting

INTRODUCTION

Surfactants (SAC's) occupy a special place among numerous classes of organic substances presenting in wastewater. The manufacturing and use of SACs in industries and in everyday life have grown dramatically during recent years with a production reaching ~ $5*10^9$ lb. alone in the USA in 1997 [1]. This trend is expected to continue in the future. Surfactants consist of three basic types: nonionic, anionic, and cationic surfactants. Of these, nonionic and cationic surfactants account for respectively 40 and 10 % of the total world surfactant production and are heavily used for industrial purposes [1]. In comparison anionic surfactants, with about 50 % of total worlds surfactant production, is primarily used for domestic detergents. Of the three types nonionic and cationic surfactants not only seems to gain heavily in production amount but also increase in complexity, whereas the production of anionic surfactants seems to be relatively stable with less complex formulations [1]. Although legislation prescribes SAC biodegradability, these cannot be completely mineralised in biological waste, treatment plants, but arrive together with their metabolites in natural resevoirs [2-4] and some are found in drinking water [3,5-10]. Chronic toxicity data of non-ionic surfactants to algae, fish and vertebrates range from 0.1 to 20 mg/l [11,12]. The ability of SACs for foam formation is a serious environmental problem, since organic contaminants and pathogenic microorganisms are concentrated in the foam, presenting an epidemiological threat. SAC's can thus cause grave disturbance and death of flora and fauna of the rivers serving as receiving waters [13]. Due to the extreme amount of SACs emitted into sewage plants and the persistence of some of the known metabolites, it is the group of environmental pollutants with the highest priority (European Council Directives 73/404, 76/464).

The creation of cheap and effective technology for the removal of SACs and other xenobiotics from water is an urgent need in modern environmental protection. Traditionally, a combination of physical, chemical and biochemical methods is applied [14]. The most ecologically acceptable approach for detoxication of xenobiotics in industrial sewage is the use of microorganisms, capable of utilising xenobiotics as their main nutritional source[15, 16]. Traditional biological purification by active sludge alone is, however, useless for highly concentrated wastewater. The content of dyestuffs can sometimes approach levels as high as 1-3 g/l, which is completely unacceptable for a biological purification process [16]. A combination of physical, chemical and biochemical methods is thus usually applied.

Recent experiences indicate that the efficiency for the purification of wastewater from toxic xenobiotics such as SAC's can be improved by combining physical-chemical with biological methods. For this purposes a unique collection of bacteria, capable of degrading a number of different xenobiotics, including SAC's, has been created [17]. During this process it was established that the process of SAC degradation is under the control of plasmids [16,18], and this knowledge enabled, the development of modern molecular genetic methods to screen for highly active strains of bacterial degraders.

Developments of physical and chemical methods of sewage water treatment also are intensive [19]. As a physical methods were recommended: treatments by electron beams [20], electrolysis [19], UV-light [21], and others. Between chemical methods it can mention addition hypochlorytes [22] hydrogen peroxide [22] and ozone [23]. Combination of these methods (UV- + H_2O_2 [23], UV- + O_3 [21]) may give a synergetic effect surpassing both separate action of each and its sum.

In this communication the results of research of SACs degradation in water solutions in process of its treatment by UV-radiation (physical component), ozone (chemical) and special bacterial culture (biotechnological component) are given. The purposes of this work is to reveal the specify of functioning of separate parts of the system in process of longtime exploitation (more than 3 months), optimisations of regimes.

Methods and Substances

As SAC's have been used: (I - NPE) - Nonylphenyldodecaetoxylateoxyl (Igepal-720 OC, Aldrich production $C_9H_{19}C_6H_4(OCH_2CH_2)_{12}OH$) and (II - LAS) - Sodium salt of dodecylphenylsulfonic acid (SIGMA production, $C_{12}H_{25}C_6H_4SO_3Na$ with the content of basic component 80%). NPE is applied as moistening and emulsioning SAC in various technological processes, especially in the textile industries, etc. It is widely widespread anionic SAC, and it is used as a base component of detergent powders, at dyeing fabrics, purification of surfaces.

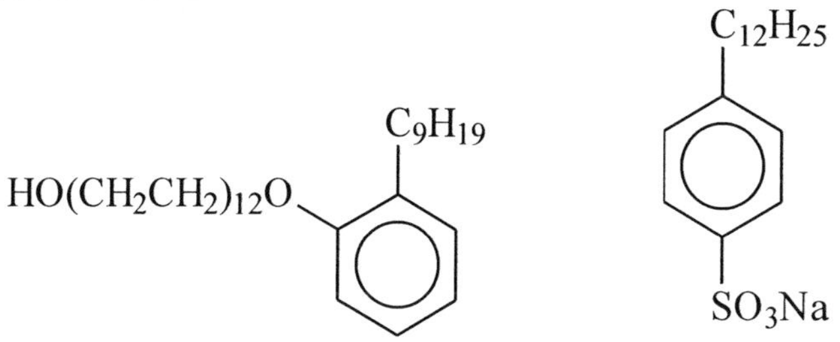

(I - NPE) (II - LAS)

The scheme of carrying out of process of photoozonization is resulted on fig 1.

As a reactor was taken glass cylinder (~1 L volume) with the hole at the bottom, inside of which there was a quark low pressure mercury lamp. In upper part of reactor the expansion has been made, allowing to collect and keep in a reactor the foam formed during ozonation. At the beginning of each experiment a stream of oxygen started to pass through all system, than the measured volume of water (1 L) was placed into reactor, then ozone generator was switched on and saturated water by ozone. After that it was included a mercury-quartz lamp and after 5 minutes pause SAC solution of the necessary concentration was injected.

Concentration of ozone was measured by light absorption in UV-range (λ=280 nm). Initial concentration was supported by a constant during all experiment.

IR-spectra of initial and ozonized SAC's were prepared on spectrophotometer Bruker IFS 48 with NaCL prism in the field of 3000 - 800 sm $^{-1}$ in $CHCl_3$ solvent or in KBr tablets.

UV-spectra of initial and ozonated water solutions of SAC's was obtained on spectrophotometer Hitachi under 210 - 300 nm. For LAS the maximal absorption was observed under 200 and 224 nm, for I - under 220, 225, 264 nm.

Reduction of concentration of SAC is registered by analytical methods. Concentration LAS determined by chloroform extraction with the subsequent measurement of optical density of painted LAS complex with methyleneblue on 650 nm.

NPE Concentration was determined on optical density of painted muddy complex NPE with Nessler reactant on 440 and 750 nm accordingly.

The toxicological control of SAC solutions has been carried out by the following biotesting methods:

1. Survival of *Ceriodaphnia dubia* (Crustacea, Cladocera) within 48 hours (acute toxicity). The toxicity coefficient was determined from the relationship between the experimental and control samples and expressed as a percentage.
2. Method of determination of water toxicity on chemotaxic reaction of infusoria. This is latest (1998) [25, 26] russian apparatus development, using the inhibition of chemotaxic reaction of infusoria (*Tetrahymena pyriformis*), when the tested sample of water contains toxic substances. The time of analysis 30-45 min.

RESULTS AND DISCUSSIONS

Ozonation and Photoozonation of LAS

In absence of light LAS reacts with O_3 slowly. At concentration of ozone 3.10^{-3}M for achievement of 25 % of conversion of 1 L LAS (250 mg/l) it is required ~5 hour. Molecules of starting LAS and it reaction products carry out a chromophore groups in they structures. It allowed to us to control the reaction dynamics by measuring increasing or decreasing of optical density of characteristic spectral bands. Changing in the UV-, and IR-spectra are represented on fig. 2 and fig. 3.

Starting II have maxims at λ = 207 and 238 nm, reaction products, probably at 218-220nm. It can be explained the smoothing of spectral curve in time of reaction and increasing of optical density at the band 218-220 on the curve 2 in comparison with 1. Decreasing of optical density at λ = 207 and 238 nm showed on destruction part of aromatic nuclei in the system, increasing band 218-220 - on formations fragments with conjugated double bonds, similar as one have in muconic acid [24]. Changing in IR-spectra stand in good correlation with all spoken above. In particularly one can see that after 30% conversion both in dark and light processes transparency of light increase in region of aromatic nuclei absorbance (1600-1500 sm^{-1}), appear bands of acid and carbonyl compounds (doublet at 1710-1720 sm^{-1}). In case of fotoozonation we saw decreasing intensity bands of C-H bonds (2700-3000 sm^{-1}), IR-spectra of dark reaction doesn't have such changing;

It leads to conclusion that alkyl substitute play an active role in fotoozonation. As a result formed a mixture of products, where represent fragments of aromatic nuclei destruction ("a" and "b" of reaction scheme), sulphoacids with short functionlized substitute ("c") as it show by the following reaction:

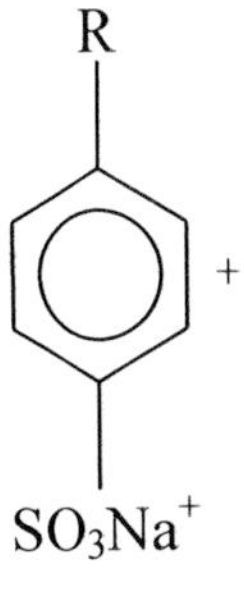

$+ O_3 \Rightarrow C_9H_{19}C(=O)OH + R'SO_3 + CH_3C(=O)C_8H_{17}C_6H_5SO_3Na$

a b c

At action of light the mechanism of reaction varies significally. It reflects both on structure of products, and on rate of process. The most probable mechanism of reaction can be presented by scheme:

$$O_3 + h\nu \rightarrow [O_3^*] \rightarrow O_2 + O$$
$$O + H_2O \rightarrow 2\ OH$$
$$HO + H\text{-}(CH_2)_{12}C_6H_4SO_3Na \Rightarrow H_2O + H\text{-}(CH_2)_{11}C^*HC_6H_4SO_3Na \Rightarrow \text{Products}$$

Ozone interacts with quantum of light, breaks up to atomic oxygen and a molecule of oxygen. Atomic oxygen reacts with a molecule of water, forming hydroxyl radical, it interacts with LAS much faster, than ozone. However its high reactionary ability causes the decrease of selectivity of its reactions and changes structure of products. From analysis of IR-spectra follows, that at photoozonization the OH-radical actively attacks alkyl substitutes of II, tearing off from them atom of hydrogen.

Ozonation and Photoozonation of NPE

Dark ozonation of NPE occurs much faster, than ozonation of LAS. At the same time it is not observed an appreciable difference in rates of ozonation and photoozonation of LAS. If to consider the potential reactionary ability of structural elements of LAS molecule towards ozone, they are standing in sequence: dodecaetoxylateoxyl >> nonyl > phenyl. In dark process ozone reacts mainly with dodecaetoxylateoxyl, shortening it in each act of reaction. At photoozonation other channels of reaction, in particular, removal of hydrogen by OH-radical from nonyl or its introduction in an aromatic nucleus open also. These changes in the mechanism are reflected by UV- and IR-specters.

In IR-spectra (fig. 4) in both cases it is observed the decrease intensity of absorption bands, characteristic for an aromatic nucleus, for the complex ethers connected with an aromatic nucleus and the appearance of a new bands, proper to keto-group. However, for photoozonated I strong intensity decreasing of bands, characteristic for alkanes which is absent in dark process, is observed. Apparently, as well as in a case with II the process of photoozonation of I is accompanied by formation of the OH-radical and its subsequent attack on alkyl radical.

Introduction of OH-group to an aromatic nucleus finds the reflection in character of changes in the UV-spectra of solutions (fig. 5) occuring during photoozonation.

One can see that in time of fotoozonation the optical density of the band 226 nm (aromatic ring) become weaker, band 214 - products of ring opening – stronger. Later 3 involve in reaction and degrade in parallel with 1. More visible the kinetics of 3 formation and degradation are shown on fig 6.

At the dark ozonation and photoozonation we have measured ozone concentrations at the inlet and outlet of reactor. This data we used for calculations of reaction rates and effective rate constants using method described earlier [28]. Obtained by us ozone consumption and values of effective rate constants are given in Table 1.

Differenes in Mechanisms of NPE Destruction at the Action of Microflora, Dark Ozonation and Photoozonization

On the scheme 1 in the simplified kind the basic stages of NPE degradation in water solutions are submitted at the action of:

a) microflorae,

b) ozone,

c) ozone and light in common.

At the action of microorganisms alongside with processes of hydrolysis of oxyethyl substitutes the nonyl end CH_3-group (which is the more stable to chemical influences including to ozone) is oxidized also. An aromatic nucleus decays at enough deep stages of hydrolysis when the intermediate product comes nearer on a structure to nonyl phenol.

At the action of ozone are attacked ethoxy-fragments near to an aromatic ring and hydroxyl group with the subsequent fragmentation of a chain of the substitute and partial or full loss by a molecule of surface-active properties. The aromatic nucleus becomes vulnerable after its activation by oxy-group.

Feature of cooperative action of light and ozone is formation in system of hydroxy-radicals which possess by much more reaction capacity, than ozone. Therefore they react with all sites of SAC molecule near to which they appeared at the moment of formation or casual diffusion.

At continious action of all types of impact the equilibrium end-products of destruction are the water and carbonic acid. Interruption of process at different stages results in mixes of products of various structure.

Toxic Properties of Products of SAC's Transformation in Result of Photoozonation and Microbial Degradation Processes

One of the primary goals of our work was improvement of optimum regimes of biodegradation. The SAC biodegradation process was carried out in the column tipe of bioreactors with immobilized degrading bacteria on inert carrier. The purification process was performed in conditions of flow cultivation. In figure 7 our installation with 2 bioreactors is represented. The top part of a reactor is filled the special inert carrier, on which a microfilm from corresponding strains microorganisms is sorbed. We varied the SAC concentration, entering to bioreactors after photoozonization process and the rate of SAC passing through reactors. It has been shown, that the combination of processes of photoozonation and biodegradation not only raise efficiency of process of cleaning, but also allow to speed up it in 2 - 3 times.

However the using of new technology of sewage treatment from polluting substances should be accompanied by the obligatory toxicological control of products of their transformation over an estimation of their influence on accepting reservoirs and water-currents. Such control has been carried out by biotesting methods.

The basic results of the toxicological control by biotesting methods of influent (water solution of SAC's, entering to reactor) without and with photoozonization and effluent (SAC's solutions, following from reactor for LAS and NPE) are represented in Tables 2 - 3. In table 2 values LC10 (IC10) are SAC concentrations at which the sample does not show toxic properties, and values LC50 are SAC concentrations at which 50 % mortality of C.d. or (IC50) - 50 % suppression of chemotaxic reaction of infusoria T.p are observed. From Table 2 is followed, that in result of photoozonation process the toxicity of LAS and NPE samples are not increase in test of C.d. mortality, and inhibition of chemotaxic reaction of infusoria T.p. decreases yet. From Table 3 is followed, that in process of microbial degradation the toxicity of LAS, NPE effluent decreses in 2-3 times in tests with C.d. and T.p.

Conclusion

This work tries to resolve two tasks - optimization for processes of SAC biodegradation with using of previous photoozonization processes and research of the influence of the photoozonization process and of the subsequent biodegradation on toxic properties of water SAC solutions. It was shown that the rate of surfactant microbial degradation after photoozone treatment increased 3-5 times. In the conditions choosen microbial cultures were effective enough to destroy more than 90% of used surfactants. The long time of continuous bioreactor exploitation has shown a good stability of microflora activity. It was shown also, that toxicological characteristics of water SAC solutions or were improved or didn't changed after processes of photoozonation and biodegradation. It were investigated kinetic characteristics of photozonation process of water SAC solutions were investigated and rate coefficients ozone-surfactants reactions were determined and their mechanisms were suggested.

This research was supported by INTAS-project 99-0995.

References

1. P.M. Morse. Soaps and detergents. *C&EN*, February 1, 35-48, 1999.
2. *Council Directive 82/242 EEC of March 31, 1982* on the approximation of the laws of the Member States relating to methods of testing degradability of non-ionic surfactants on amending Directive 73/404/EEC
3. J. Rivera, D. Fraisse, F. Ventura, J. Caixach, A. Figueras, Z. Fresenius. *Anal. Chem.* **328** (1987) 577.
4. M. Ahel, W. Giger, C. Schaffner, *Water Research,* **28** (1994) 1143-1152
5. J. Rivera, F. Ventura, J. Caixach, M. de Torres, A. Figueras and J. Guardiola, *Intern. J. Environ. Anal. Chem.,* **29** (1987) 15.
6. F. Ventura, D. Fraisse, J. Caixach, J. Rivera. *Anal. Chem.,* **63** (1991) 2095.
7. L.B. Clark, R.T. Tosen, T.G. Hartmann, J.B. Louis, I.H. Suffet, R.L. Lippincott, J.D. Rosen, *Intern. J. Environ. Anal. Chem.,* **47** (1992) 167.
8. H. Fr. Schöder, *J. Chromatogr.,* **554** (1991) 251.
9. H. Fr. Schroder, *Wat. Sci. Technol.,* **25** (1992) 241.
10. H. Fr. Schroder, *J. Chromatogr.,* **643** (1993) 145.
11. M.A. Lewis. Ecotoxicol. *Environ. Safe,* **20** (1990) 123.
12. M.A. Lewis, *Wat. Res.,* **25** (1991) 101.
13. S. Stavskaya, V. Udod, L.Taranova, I.Krivez. *"Naukova dumka"*, Kiev, 1988. p18. (in Russian).
14. N.A. Klimenko, M.N. Timoshenco. *Water Chemistry and Technology*, 1993, V.15, N7-8, P. 534-566.
15. A.M. Koganovskii, N.A. Klimenkio. *Physical-chemical methods of industrial waste water purification from surfactants*. Kiev, Naukova Dumka, 1974
16. L.A. Taranova. *Ecological aspects of bacterial destruction of ionogenic surfactants, Author's abstract of the doctorate*, Dnepropetrovsk, 1994 (in Ukraine).
17. S.S. Stavskaja, L.A.Taranova, I.A. Krivetz. *Avt. Svid. USSR, №962304, Bull. Izobr.* 1982, **№36**, p.12.

18. A.N.Dubeikovskii, L.A.Taranova et.al. *Microbiology*, vol.66, N3. 1997, pp.316-319.
19. A.Ya. Efimov, I.M. Tavartkiladze, L.I. Tkachenko. *Purification of textile industry waste waters*. Kiev, "Technica", 1985, 232 p.
20. E.A. Podzorova. *Dr.Sci. Dissert. Thesis*. Moscow, Inst. Phys.Chem. Russ. Acad. Sci., 2001.
21. J.L. Matteman, H. Grootendorst. *Proceedings of 10th Ozone World Congress*, Monaco, 1991, v.2, p.237.
22. M.A. Shevchenko. *Oxidants in the Water Treatment Technologies*. Kiev, Naukova Dumka, 1979.
23. S.D. Razumovskii. *Zhurn. Mendeleev.Chim.Soc.USSR*, 1990, v.35, p.77.
24. United States Environmental Protection Agency Method 1002.2. Oct.11.95 1034. No 001 P 03 *Cladocera, Ceriodaphnia dubia,* survival and reproduction test method.
25. Toxicological methods of control. Method of determination of water toxicity on chemotaxic reaction of infusoria. *PND F T 14.1:2:3:4.2-98*. Russian Federation State Committee of Environmental Protection. Moscow, 1998. 14p. (in Russian).
26. Methodical instructions on using of infusoria *Tetrahymena pyriformis* as a test-culture in device "Biotester-2". Ministry of agriculture of Russian Federation, Moscow, 2000 (in russian).
27. M.L. Konstantinova, S.D. Razumovskii, G.E. Zaikov. *Bull. Acad. Sci. USSR*, 1991, p. 324
28. S.D. Razumovskii, G.E. Zaikov. *Ozone and its reaction with organic compounds*. Elsevier, Amsterdam, Oxford, New York, Tokyo, 1984.

Chapter 6

Evaluation of Individual Organism Sensitivity to Carcinogenic Action

***V.S. Sibirtsev* [1 #], *M.L. Tyndyk* [2] *and A.V. Garabadzhiu* [3]**

[1] Mendeleyev Russian Research Institute for Metrology St. Petersburg, 190005 Russia;

[2] N.N.Petrov Research Institute of Oncology, Russian Ministry of Public Health, St. Petersburg, Russia;

[3] St. Petersburg State Technological Institute, St. Petersburg, 190013 Russia.

Abstract

The dynamics of excretion from organism and deactivation of the benzo[a]pyrene (*BP – one of the most widespread and dangerous among carcinogenes*) is investigated on the example of rats. Is shown, that duration of rat life after beginnings of the BP inject correlates with a content (±)-trans-7,8-dihydroxy-7,8-dihydrobenzo[a]pyrene in animal urine (*were determined on specially techniques developed by us*). Besides is shown, that the individual sensitivity to carcinogen in case of single BP inject is mainly determined by efficiency of excretion of the BP active forms from organisms, whereas after multiple BP inject in organism it is determined by mechanisms of BP enzyme deactivation.

Key words: carcinogenes, benzo[a]pyrene

Benzo[a]pyrene (BP) is one of the most widespread and dangerous among carcinogenes. It will be formed, in particular, as a result of partial combustion of various organic substances. However, the individual sensitivity of the people to BP (*as well as to other carcinogenes*) varies. [1-3].

For revealing of this mechanism, we took a group of rats, whose BP was entered in the form of a water-lipid emulsion unitary (*in dozes 100 or 200 mg/kg*) or repeatedly (*on 10 mg/kg – 10 time with a interval of 10 days*). As far as BP able to render on organism not only carcinogenic, but also all-toxicy action, as parameter, describing individual sensitivity of rats

[#] To whom correspondence should be addressed; fax: +7 (812) 327-97-76;
e-mail: vs1969r@mail.ru; websyte: http://www.vs1969r.narod.ru/indexen.htm.

to presence BP, we chose the common life duration of the animals after beginnings of BP inject. This parameter was compared to a total output of (±)-trans-7,8-dihydroxy-7,8-dihydrobenzo[a]pyrene (7,8-BP – *main carcinogen BP metabolite*) in animal urine, what was canvassed during the first 5 days after each BP inject to him and analyzed on specially techniques developed by us [4-7]. Thus, the term – 5 days was chosen, as far as (*as was established during preliminary researches*) for this time is removed the main quantity of BP, after injecting in rats.

The results of conducted researches are shown on fig.1-4. Of these data it is possible to make following conclusions.

1. For BP (*outside of dependence on a mode of its inject in organism*) takes place about equal contribution of carcinogenic and all-toxicy effects in reduction of rat life duration.
2. The mode of multiple BP inject in organism is more sparing in comparison with a mode of single BP inject in the same total doze because in the latter case a direct excretion of the BP active forms from organism will be realized only, whereas in the first of considered cases organism "includes" addition mechanismes of adaptation (*such, for example, as BP enzyme deactivation*).
3. Duration of rats life after beginnings of the BP inject correlates (*with a sufficient degree of reliability*) with a content (±)-trans-7,8-dihydroxy-7,8-dihydrobenzo [a]pyrene in animal urine.

These conclusions can be are extrapolated to peoples as well. (*taking into account of course the distinctions in speed metabolism of people and rats, etc*). It, in particular, can serve the basis for development of a technique enabling most subject to the negative effect of carcinogenes, and thus to have a opportunity to apply under the attitude to them appropriate measures.

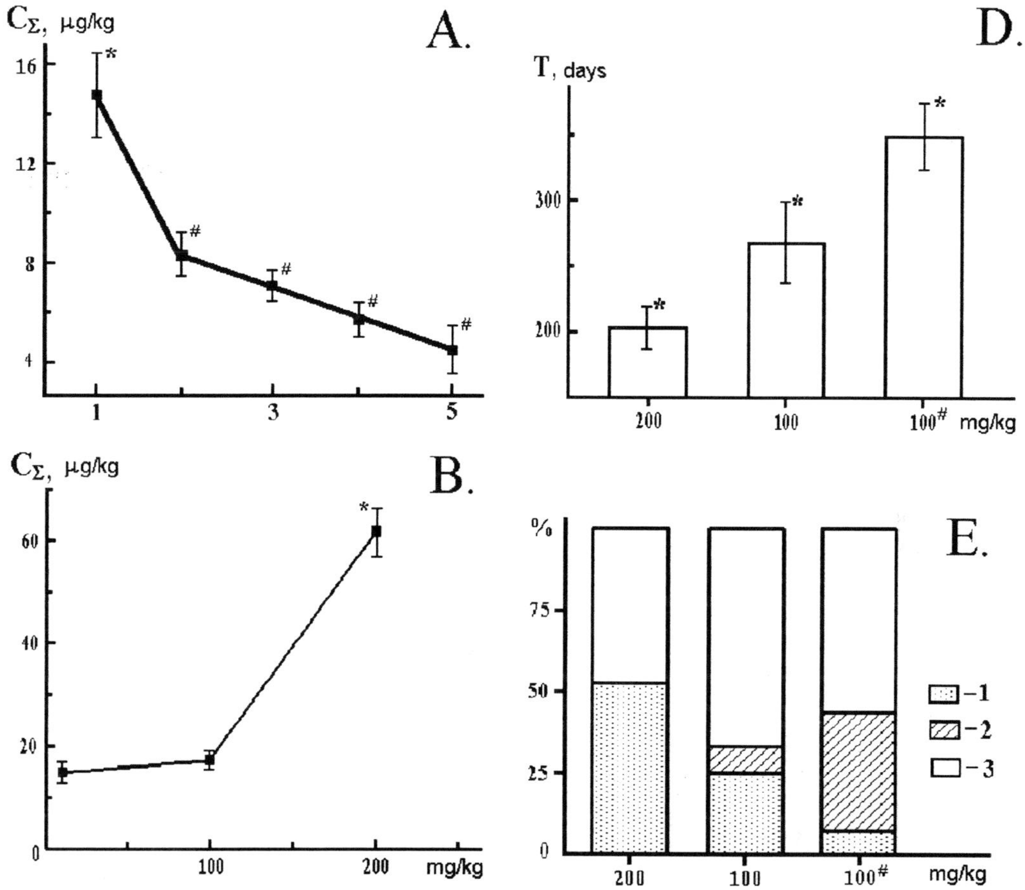

Fig.1. Change of average on group parameters "
depending on a doze and mode benzo[a]pyrene (BP) inject.

On abscissa axis are shown: **A** – number of injects, **B,D,E** – total doze of BP inject; on ordinate axis are shown: **A,B** – average on group total content of (±)-trans-7,8-dihydroxy-7,8-dihydrobenzo[a]pyrene (7,8-BP) in animals urine at the first 5 days after BP inject to him; **D** – average on group duration of rat life; **E** – percent of animals in group, at which after BP inject internal bodies tumpures were formed (1), underskin tumpures were formed (2) or tumpures were not formed (3).

The symbol "#" marks a mode of multiple BP inject in organism (*on 10 mg/kg with a interval of 10 days*); symbol "*" – marks a significances of parameter, authentically ($p < 0.05$) distinguished from next with them of significances.

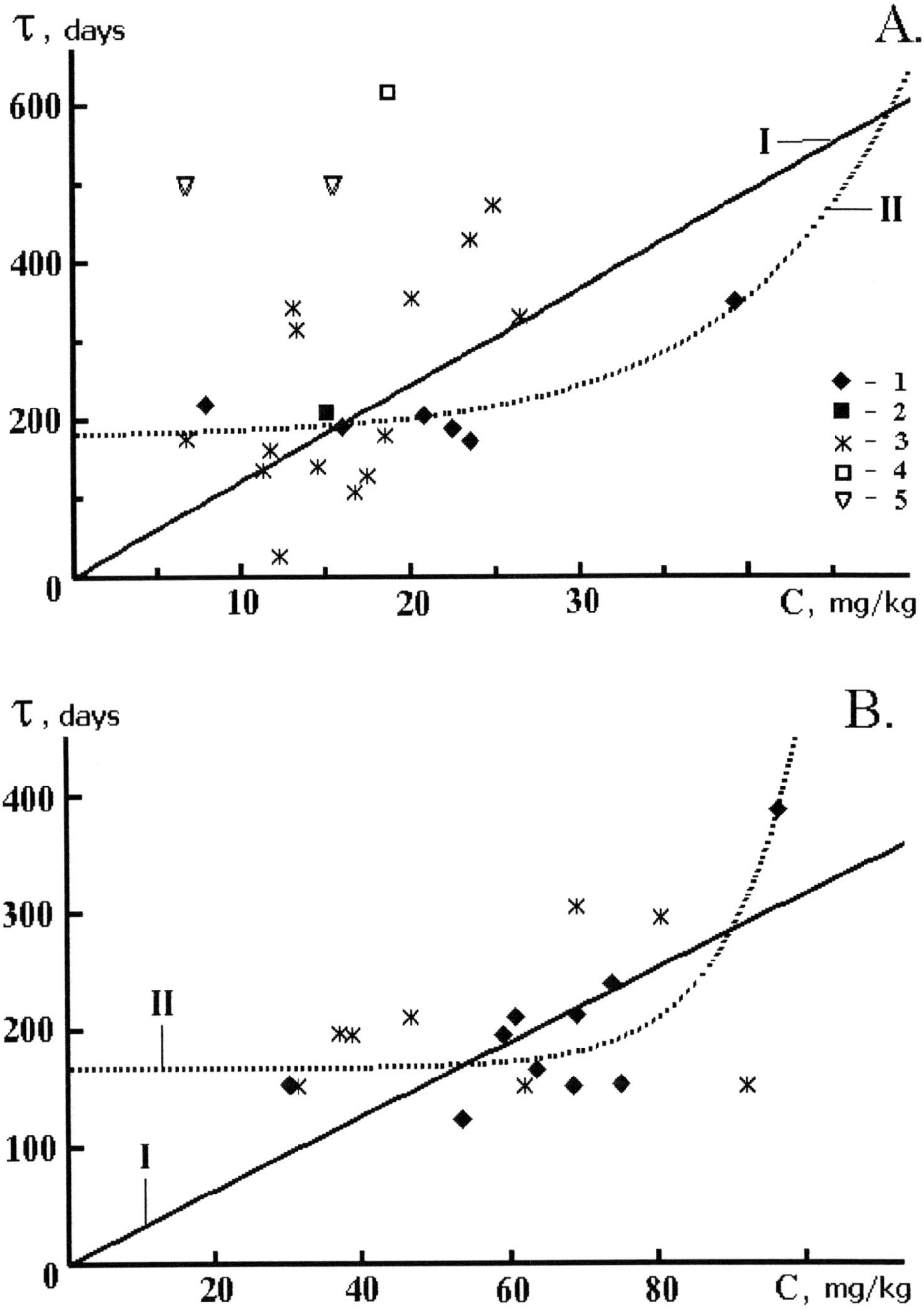

Fig.2. Dependence of individual rat life duration (τ) from total content 7,8-BP in animals urine (**C**) at the first 5 days after single BP inject to him in a doze 100 (A) and 200 (B) mg/kg.

Conditional designations on figure: the points **1-3** designate experimental data for the animals, at which after the BP inject: **1** – internal bodies tumpures were formed, **2** – underskin tumpures were formed, **3** – tumpures were not formed; the points 4 and 5 designate data, which were eliminated pursuant to a technique, described in paper [7], for 2 and 3 animals groups, accordingly.

The lines designate regression equations were calculated for: **I** – all researched animals group (fig.**A** – $\tau = 12{,}2^{x}$**C**, *correlation factor* **r**=0,510, *reliability* **p**<0,05; fig.**B** – $\tau = 3{,}15^{x}$**C**, **r**=0,484, **p**<0,05); **II** – animals, at which after the BP inject internal bodies tumpures were formed (fig.**A** – $\tau = 178 + 3{,}23^x e^{C/10}$, **r**=0,917, **p**<0,01; fig.**B** – $\tau = 167 + 0{,}014^x e^{C/10}$, **r**=0,893, **p**<0,01).

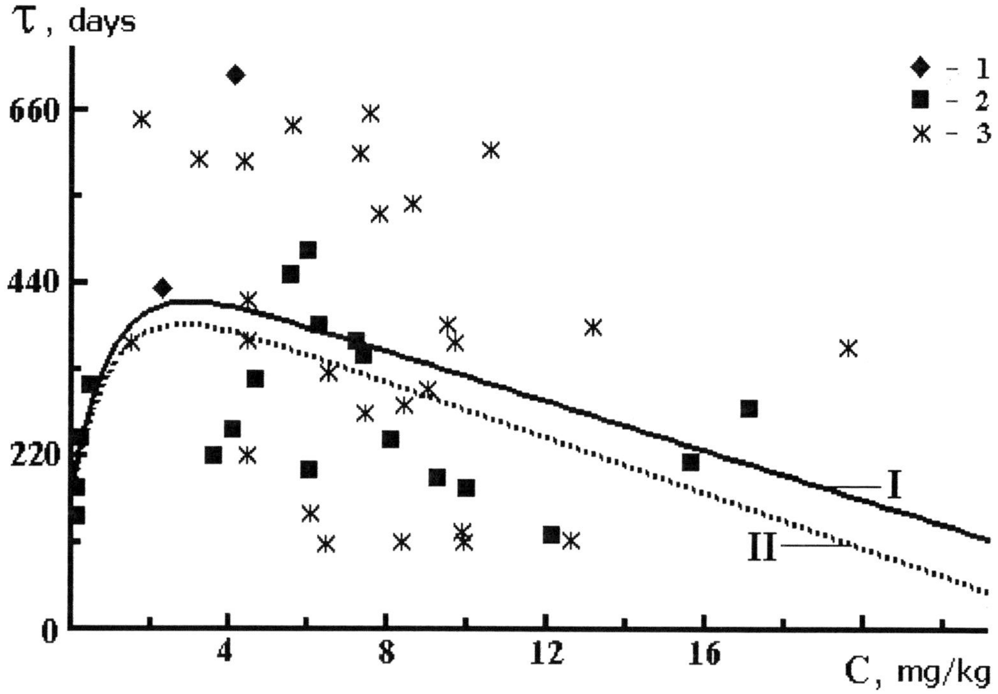

Fig.3. Dependence of individual rat life duration (τ) from total content 7,8-BP in animals urine (**C**) at the first 5 days after 3 inject at the multiple BP inject to him (*in a doze 10 mg/kg – 10 time with a interval of 10 days*).

Conditional designations on the schedule for points – same, as on fig.2.

The lines designate regression equations were calculated for: **I** – all researched animals group ($\tau = 477 - 15{,}6^x C - 309^x e^{-C}$, **r**=0,372, **p**<0,05); **II** – animals, at which after the BP inject internal bodies tumpures were formed ($\tau = 454 - 17{,}5^x C - 296^x e^{-C}$, **r**=0,547, **p**<0,05).

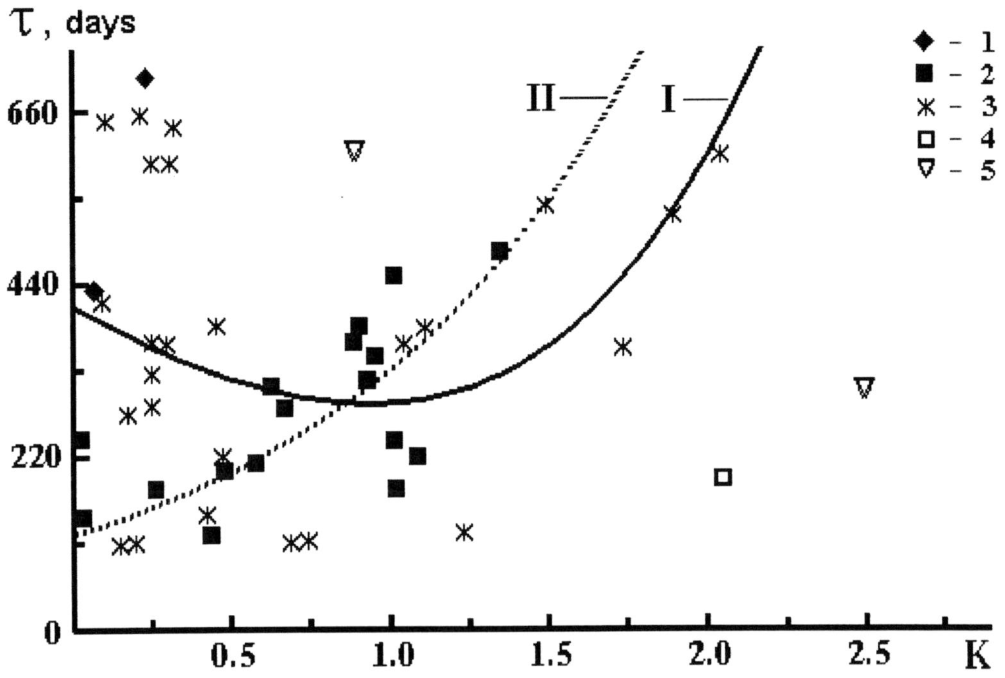

Fig.4. Dependence of individual rat life duration (τ) from relation of total contents 7,8-BP in him urine at the first 5 days after 3 and 1-й injects (K) at the multiple BP inject to animal organism (*in a doze 10 mg/kg – 10 time with a interval of 10 days*).

Conditional designations on the schedule for points – same, as on fig.2.

The lines designate regression equations were calculated for: **I** – all researched animals group ($\tau = 265 - 377^{x}$**K** + 147^{x}**e**^{K}, r=0,392, p<0,05); **II** – animals, at which after the BP inject internal bodies tumpures were formed ($\tau = 121^{x}$**e**^{K}, r=0,663, p<0,01).

References

1. IARC Monographs on the Evaluation of Carcinogenic Risk to Humans. *Supp.* 7, IARC (Lyon, France), 1987.
2. A. J. Likhachev, D. Sh. Beniashvili, V. J. Bykov et. all: Environ. *Health Perspecti.* **98**, 211.
3. A. J. Likhachev, M. A. Zabezhinskij, V. J. Bykov et. all: *Cancer Letters.* **78**, 163.
4. I. V. Savochkina, P. P. Dikun, M. A. Zabezhinskij et. all: *Questions of Oncology,* **36**, 1221 *(in Russian).*
5. A. J. Likhachev, M. A. Zabezhinskij, P. P. Dikun et. all: *Questions of Oncology,* **38**, 950 *(in Russian).*
6. M. A. Zabezhinskij, A. J. Likhachev, V. J. Bykov et. all: *Experimental Oncology,* **16**, 431 *(in Russian).*
7. V. S. Sibirtsev, M. L. Tyndyk, S. D. Ivanov: *Questions of Oncology,* **46**, 594 *(in Russian).*

Chapter 7

EXPRESS–METHOD OF THE GENERAL EVALUATION OF MICROBIOLOGICAL POLLUTION OF WATER BY FLUORESCENT DNA–SPECIFIC DYES

V.S. Sibirtsev [1 #] ***and A.V. Garabadzhiu*** [2]
[1] Mendeleyev Russian Research Institute for Metrology;
St. Petersburg, 190005 Russia
[2] St. Petersburg State Technological Institute,
St. Petersburg, 190013 Russia

ABSTRACT

A new method is described which can be applied for express valuation of the level microbial pollution of water directly on the place where the sample is taken.

Key words: microbial pollution, fluorescent dyes, DNA.

Microbiological pollution is one of important parameters of quality for a water, food products, etc. The most standard method for determination of it is plating of analyzed material on a medium with subsequent (*after incubation*) estimation of the number microbial colones formed. The main defects of the given method are: significant time necessary for analysis fulfilment (from *24 hours and more*), as well as difficulty for automation of the analysis process, and accordingly, significant subjective factor.

As alternative way of valuation of the general microbiological pollution in water we have developed, in comparison to the above-mentioned standard method, has smaller sensitivity, but requires considerably smaller times for analysis fulfilment (*within several minutes*). Besides our method has considerably good possibilities for automation of the analysis process. It also differs in that analysis (*when available portable fluorimeter which are*

[#] To whom correspondence should be addressed; fax: +7 (812) 327-97-76 *manufactured by industry*) can be executed directly at places where the sample is taken, without transporting them to a stationary laboratory.

The essence of the presented method consists of the following. To the taken sample add a water solution ***A***, containing: 2M ***NaCl*** + 0.1M ***Na_2EDTA*** (*ethylenediaminetetraacetic acid disodium salt*) + 0.01M ***Tris*** (*2-amino-2-hydroxymethyl-1,3-propandiol*) + 0.5 vol. % ***Triton X-100*** *(union detergent)* [1] – which enables to support the sample pH in a range 7.8–8.2, as well as destroys cell membranes for maintenance unobstructed access of dye to microbial DNA. Thus the volumetric ratio of initial sample to a solution ***A*** should be, about, 10 to 1. Further, samples are incubated during several minutes (*at* $\mathbf{t^0}$=25 ^{0}C, *up to a establishment of the stable indications on fluorimeter*). Then, the values of $\mathbf{I_o}$ parameter (*fluorescent intensity of modified sample*) are determined at lengths of waves of excitation and emission: 360 and 460 nm, accordingly. Then, to the modified sample add a solution of the DAPI (*4',6-diamidino-2-phenyl-indole – fluorescent DNA–specific dye*) [2,3] and under the same conditions, as earlier, determined the value of the $\mathbf{I_d}$ parameter.

Then, value: $\mathbf{K} = \mathbf{lg[I_o/(I_d - I_o)]}$ is calculated and under the formula: $\mathbf{C} = \mathbf{10^{a_0 + a_1 K}}$ (where the factors a_0 and a_1 find experimentally for a particular fluorimeter type, modes and conditions of measurement) are determine the total quantity of microbes in sample (**C**).

With using the TKO-100 portable microfluorimeter (manufactured by the firm "Serva", Germany) the given technique has allowed us for various microbes strains (Escherichia coli, Salmonella, Shigella lexneri and other) to determine the total quantity of microbes in samples from 20 cells/ml, with a error of not more than 30 %, expending not more than 10 minutes on one measurement.

Our method can be used, in particular, for express valuation of level microbial pollution directly at the place where the sample is taken. Then (if a situation is critical or is near that), the refinement of pollution parameters can be conducted in a stationary laboratory.

The base level of microbe detection by our method can be reduced with the use of a flow citofluorimeter. Thus the method can be applied only in stationary conditions. Besides our method can be advanced the immobilization the fluorescent dye on a firm surface, use of several dyes with various properties, application of the "internal standard" for reduction of a determination error, etc.

More information on our work and other adjacent directions can be found on our website: http:\\www.vs1969r.narod.ru\indexen.htm.

REFERENCES

1. P. R. Cook: *EMBO Journal*, **3**, 1837 (1984).
2. M. L. Barcellona, R. Favilla, J. Von Berger et. all: *Archives Biochem. Biophys.*, **250** (1), 48 (1986).
3. V. S. Sibirtsev, A. V. Garabadzhiu, S. D. Ivanov: *Russian Journal of Bioorganic Chemistry*, **20** (6), 548 (1994).

Chapter 8

QUALITATIVE AND QUANTITATIVE ANALYSIS OF NUCLEIC ACIDS BY TWO FLUORESCENT DYE SYSTEMS

***V.S. Sibirtsev* [1 #] *and A.V. Garabadzhiu* [2]**

[1] Mendeleyev Russian Research Institute for Metrology;
St. Petersburg, 190005 Russia

[2] St. Petersburg State Technological Institute,
St. Petersburg, 190013 Russia

ABSTRACT

Opportunities for application for analysis for total quantity of DNA in sample, as in the structure from two commercial fluorescent dyes (*ethidium bromide and Hoechst-33258*), having by similar spectral and binding properties are considered.

Key words: DNA analysis, fluorescence dyes, spectral analysis.

The analysis of geno structure and quantity DNA in a sample is a topical problem in many areas of biotechnology. Man-made synthetic fluorescent nucleic–specific dyes are presently used in this regard. Advantages of the given method are: its simplicity, express sensibility to small quantities of DNA in the sample. However, it does not have high selectivity. The analysis of DNA structure in the sample is also only slightly informative.

For elimination of above-stated defects we used for the DNA analysis a system from two widespread commercial dyes: ethidium bromide (**EB**, *2,7-diamino-10-ethyl-9-phenyl-phenantridiumbromid*) and Hoechst–33258 [**HT**, *2'-(4-hydroxyphenyl)-5-(4-methyl-1-piperazinyl)-2,5'-bis(1H-benzimidazole)*] (*look at the scheme*), earlier used for decision similar problems only on separateness [1-6].

[#] To whom correspondence should be addressed; fax: +7 (812) 327-97-76; e-mail: vs1969r@mail.ru; websyte: http://www.vs1969r.narod.ru/indexen.htm.

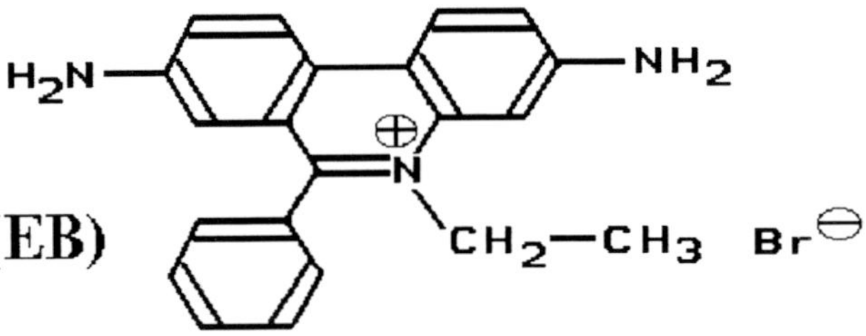

(EB)

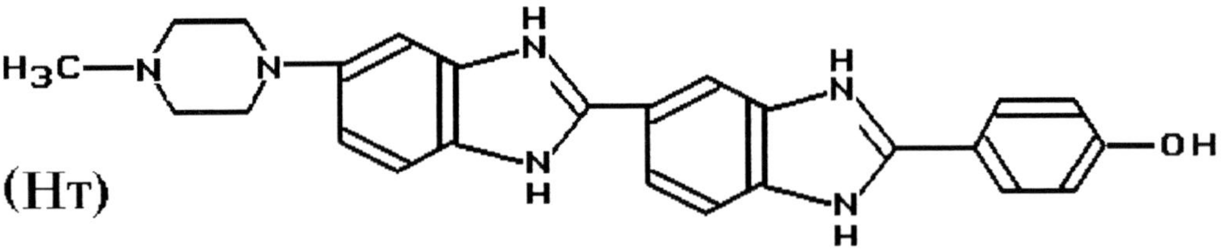

(HT)

Scheme.

Thus was marked, that the spectral properties of a considered system are determined by interaction a included in its dyes as with nucleic acid, as with one another. In particular, at joint sorption **EB** and **HT** (*with wave lengths of maxima fluorescent excitation in the visible area of a spectrum: 520 and 350 nm – and emission: 600 and 450 nm – accordingly; look the fig.1*) on DNA the fluorescent resonanse transfer of energy (FRET) can take place between molecules of this dyes: the phenomenon, when at **HT** (*donor*) fluorescention excitation to glow begin not only its molecules, but also **EB** (*acceptor*) molecules, that sorption on polynucleotid on distance from **HT** molecules a smaller emission wave length of the last (*look the fig.2*); and the efficiency of such transfer depends as on distance between molecules of the donor and acceptor of energy, as on the size of the intersection area of spectra of emission of the donor and fluorescence excitation of acceptor [7-11].

Besides it is known, that as interaction with polynucleotid the **HT** is externally binding compound, preferablly specific to sites of polynucleotid double helixe, containing by three consistently located AT- and one GC- nucleotid pairs; and **EB** is intercalator, which is built between two nucleotid pairs on substrat double helixe with some preference to its GC-rich sites, and in a considerably greater degree, than **HT**, specific to the higher orders of a polynucleotid structure [2-6].

Thus, there are the basis to believe, that sharing of its dyes is promised additional effects, presenting interests in frameworks structure-functional study of DNA: down to a opportunity of express valuation of a degree superhelixe polynucleotid in genom and even lengths of it telomer sites [12-15]. For that, as one of probable schemes of the polynucleotid analysis, it is possible to offer, following.

1. We add to 0-sample a special buffer solution and lytic mix (for maintenance as pH in a system at a required level, as best access of dye molecules to DNA) and divide it on two parts (1- and 2-sample).

2. To the 1-sample we add a **EB** and on it fluorescence change (relating to background significance - without sample - at excitation and emission waves lengths: 520 and 605 nm) determine the total DNA concentration in sample (**C_{DE}**).
3. The 2-sample we saturate with a **HT** and on it fluorescence change (at excitation and emission waves lengths: 350 and 455 nm) also determine the total DNA concentration in sample (**C_{DH}**).
4. Then to the 2-sample we add a **EB** (which will take on a substrate sites, not yet connected with **HT**). Then on system fluorescence change at excitation and emission waves lengths: 350 and 605 nm, we evaluate the efficiency of fluorescent resonant energy transfer (FRET) from **HT** to **EB** (**K**), which should correlate with number of sites on DNA, boundary between AT- and GC-rich polymer areas, the reasonably significant part from which, obviously, should have on telomer sites (having by a structure: [TTAGGG]n – for vertebrates and human).
5. We calculate a factor: **$S=C_{DE}/C_{DH}$**, the increase of which should, in a certain measure, to reflect reduction of a degree sample DNA superhelixe.

The conditions, when the significances of **C_{DE}** and **C_{DH}** will be equal, can be achieved, for example, by means of preliminary processing of the sample by ultrasound *(for homogenization DNA)*. In case the directly measurements of vital DNA the more adequate information about it total quantity in sample are given by value of **C_{DE}**, though the value of **C_{DH}** has large sensitivity

The DNA analysis scheme, which was described above, was tested by us, in particular, on two groups of rats on 10 animals in each; selected such, that the I-group included the animals with average age – 2 months and weight – 140 gr, and II-group included the animals with average age – 1.5 year and weight – 350 gr. As a result, us were received statistically authentically ($p<0.01$) distinguished significances of parameters **S** and **K**, middling on I and II groups – that and are followed to expect, as far as with age *(and distinction in "speeds of gerontology processes" for rat and men make, about, 20:1)* condition of a genetic material in cells should be worsened, in particular, at the expense of reduction of a degree DNA superhelixe, as well as length telomer sites of genome.

More information on this work and other adjacent directions can be found on the website: http:\\www.vs1969r.narod.ru\indexen.htm.

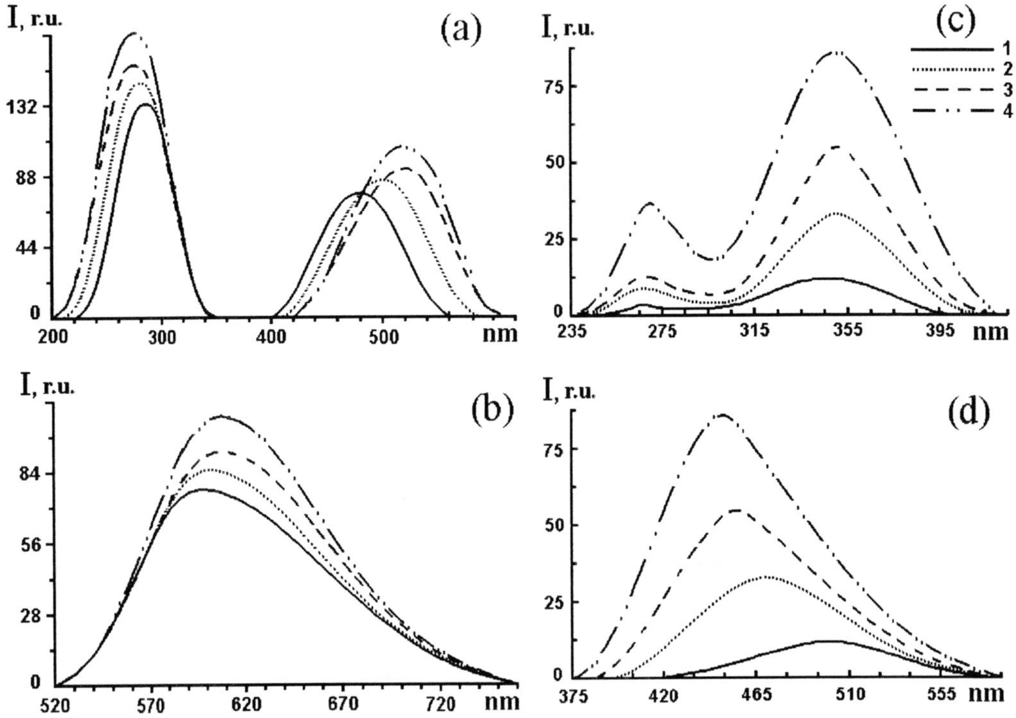

Fig.1. Spectra of fluorescence excitation (**a**,**c**) and emission(**в**,**d**) a dyes **EB** (**a**,**b**) and **Ht** (**c**,**d**) at the presence of various quantity DNA (*from calf timus*) in aqueous buffer *A* containing 0.01 mol/l of NaCl, 0.01 mol/l of Na_2EDTA (*ethylenediaminetetraacetic acid disodium salt*) and 0.01 mol/l of Tris (*2-amino-2-hydroxymethyl-1,3-propandiol*).

Spectra were recorded on "Hitachi" model 850 spectrofluorimeter (*Japan*) at slit width 3 nm, scan rate 120 nm/min, response time 2 s and normal amplification of photoelectron multiplier. Thus standard 1-cm square-section cells were used. All used reagents were received from the firm "Serva" (*Germany*).

Spectra were registered at fixed emission waves lengths: 600 (**a**) and 450 (**c**) nm – and excitation waves lengths: 520 (**b**) and 350 (**d**) nm. Curve 1-4 there correspond the relation of moles concentration C_{DNA}/C_{dye}= 0, 5, 10 and 25. $C_{EB}=8\times10^{-6}$ mol/l and $C_{Ht}=1\times10^{-6}$ mol/l.

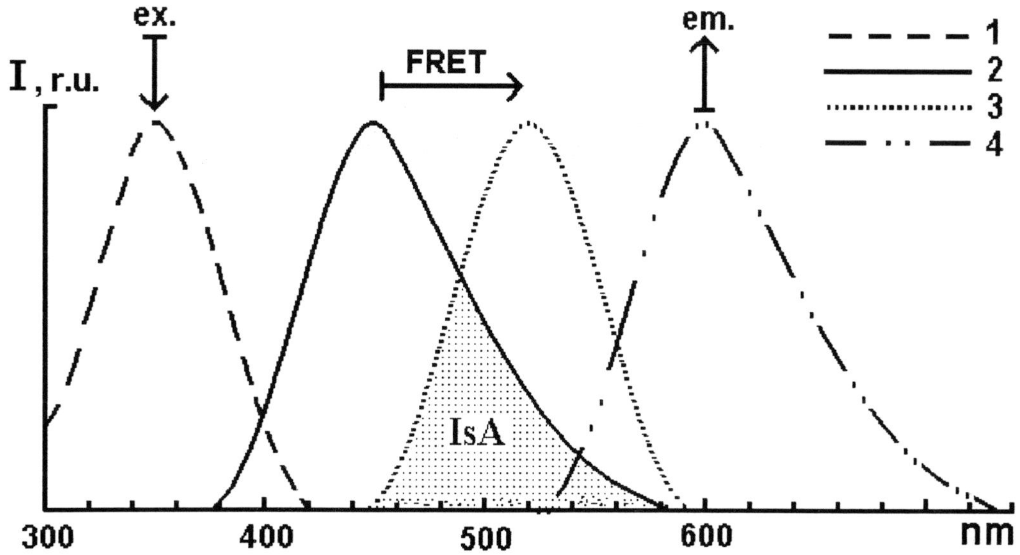

Fig.2. Scheme of the fluorescent resonanse transfer of energy (FRET), occuring between **HT** and **EB** in presence DNA.

Curves **1** and **2** are designated the spectra of fluorescent excitation (*at a emission wave length – 450 nm*) and emission (*at a excitation wave length – 350 nm*), which are registered for **HT** in presence DNA. Curves **3** and **4** are designated the spectra of fluorescent excitation (*at a emission wave length – 600 nm*) and emission (*at a excitation wave length – 520 nm*), which are registered for **EB** in presence DNA. As "**IsA**" is designated the intersection area of **HT** emission and **EB** fluorescent excitation spectra, in which can occur FRET, in case when the **HT** and **EB** molecules are located close enough from each other; "**ex.**" – excitation; "**em.**" – emission.

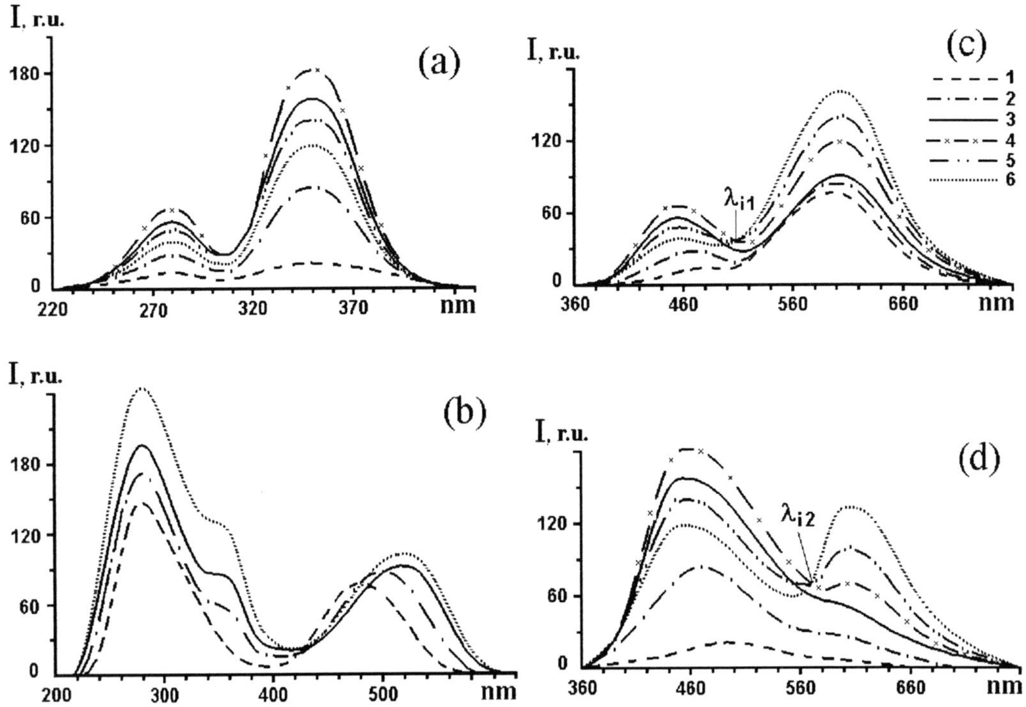

Fig.3. Fluorescence spectra of a system "**HT+EB**" (**C_{EB}/C_{HT}**=8) at the presence of various quantity DNA, registered at fixed emission waves lengths: 450 (**a**) and 605 (**b**) nm – and excitation waves lengths: 280 (**c**) and 350 (**d**) nm.

Curve 1-6 there correspond the relation of moles concentration **C_{DNA}/C_{HT}**= 0, 5, 10, 15, 20 and 25. As "λ_{i1}" and "λ_{i2}" are designated izobestic points present on fig.**c** and fig.**d**, accordingly. Others conditions of measurements, reagents and etc. – as on fig.1.

REFERENCES

1. C. F. Cesarone, C. Bolognesi, L. Santi: *Anal. Biochem.* **100** (1), 188 (1979).
2. A. R. Morgan, J. S. Lee, D. E. Pulleyblank et. all: *Nucl. Acids Res.* **7**, 547 (1979).
3. P. E. Pjura, K. Grzeskowiuk, R. E. Dickerson: *Journal of Molecular Biology.* **197**, 257 (1987).
4. V. S. Sibirtsev, A. V. Garabadzhiu, S. D. Ivanov: *Russian Journal of Bioorganic Chemistry.* **20** (6), 605 (1994).
5. 5. I. Haq, J. E. Ladbury, B. Z. Chowdhry et. all: *Journal of Molecular Biology.* **271**, 244 (1997).
6. O. F. Borisova, A. K. Schelkina, A. T. Karapetjan et. all: *Russian Journal of Molecular Biology.* **32** (5), 855 (1998).
7. V. L. Ermolaev, E. N. Bodunov, E. B. Sveschnikova et. all: *Nonradiative excitation energy transfer*. Nauka (Leningrad), 1977. (*in Russian*).
8. L. Stryer: *Annu. Rev. Biochem.* **47**, 819 (1978).
9. R. M. Clegg: *Methods Enzymol.* **211**, 353 (1992).
10. S. Speiser: *Chem. Rev.* **96**, 1953 (1996).

11. L. Lankiewicz, J. Malika, W. Wiczk: *Acta Biochimica Polonica.* **44**, 477 (1997).
12. A. K. Miker, D. S. Koffi: *Russian Journal of Biochemistry.* **62** (11), 1547 (1997).
13. E. V. Kurenova, D. M. Mejson: *Russian Journal of Biochemistry.* **62** (11), 1453 (1997).
14. V. P. Skulachev: *Russian Journal of Biochemistry.* **62** (11), 1394 (1997).
15. E. X. Blekbern: *Russian Journal of Biochemistry.* **62** (11), 1400 (1997).

Chapter 9

DEVELOPMENT OF NONLETHAL METHODS OF RODENT POPULATION CONTROL

V.V. Voznessenskaya,[1] S.V. Naidenko,[1] L. Clark [2] and D.S. Pavlov [1]
[1] A.N. Severtzov Institute of Ecology & Evolution RAS, Moscow, 119071,Russia,
[2] National Wildlife Research Center, United States Department of Agriculture, Fort Collins, CO 80521, USA

ABSTRACT

Several recent surveys indicate that rodents have the largest negative impact amongst vertebrates on agricultural production. In addition to agricultural losses, human health and safety from rodent transmissible zoonoses are of concern We examined the influence of predator chemical cues derived from predators on reproductive output of rats, mice and voles. Animals responded to predator chemical cues with suppressed reproductive output. A number of active compounds from predator scents have been detected and tested under laboratory and semi-natural conditions.

Key Words: Rodents, reproduction, population control, nontoxic repellency, reproductive inhibitors

INTRODUCTION

Rodents cause considerable economic damage to field and fruit crops on an annual basis.

Several recent surveys indicate that rodents have the largest negative impact amongst vertebrates on agricultural production. In addition to agricultural losses, human health and safety from rodent transmissible zoonoses are of concern. The magnitude of the problem is exemplified as a single commodity, apples. In the United States of America (USA) alone, rodents cause $90 million of damage each year (apples alone are a $1.8 billion industry). Efforts to control rodents responsible for apple crop damage include the application of 321 000 pounds of herbicides for habitat management and 1 000 pounds of acute rodenticides including strychnine, zinc phosphide and chlorophacinone (National Agricultural Statistical

Service). Agricultural losses and zoonoses attributable to rodents are exacerbated during periods where their population erupt. Methods that can dampen these irruptive population cycles would prove highly desirable. It is our goal to develop a product that will dampen the amplitude of these rodent population cycles. Reproductive control in wildlife species who come into conflict with humans has received increasing attention as a humane method for managing wildlife populations. Moreover, modeling studies show that contraception as a tool to management populations is best suited for species with high population turnover, i.e., short generation time and high reproductive output. Thus, rodents are ideal targets for this management tactic. Predator urine is used as a wildlife management tool to repel herbivorous animals from areas and to protect horticultural plantings from being eaten. Fundamentally, avoidance of predator urine by potential prey, and by implication the areas where predators frequent, is presumably evolutionary advantageous because it lowers the risk of predation. Potential prey can discriminate predator urines as opposed to that of other herbivores on the basis of the urine's odor.

One consequence of a high meat diet is the presence of sulfurous compounds in the urine. These compounds result from protein digestion and metabolism. When sulfurous compounds are removed from predator urines by mercury treatment herbivorous animals are no longer repelled by the urine's odor (Nolte et al., 1994). While studying the effects of predator odors on various aspects of rodent behavior we noted that reproductive behavior and output were affected as well (Voznessenskaya et al., 1992; 2000, Voznessenskaya, 2003). We now present evidence to support this observation.

MATERIALS AND METHODS

Test Subjects

Test subjects were 3-4 month old Norway rats (Rattus norvegicus) and 4-6 month mice (Mus musculus); both from an outbred laboratory population. Before the start of the experiments, females were housed in groups of 3-4, and males were housed singly. Experimental rooms were illuminated on 14:10 hours light:dark schedule, and maintained at 20°C. Food and tap water were provided *at libitum*. Virgin females in proestrus/estrus, as determined by vaginal cytology, were chosen for the mating experiments. Sexually experienced males that were not mated in the 14 days before the test were used as sires. The morning after pairing, the females were checked for successful mating, as indicated by the presence of a vaginal plug. Successfully mated females were then housed singly.

Reproductive Output

For each experimental group, the total number of offspring was counted as well as number of pups per female; sex ratio was determined. In addition, we also measured the gestation period.

Determination of Implantation Sites in Utero and Postimplantation Loss

Females were necropsied 2-3 days after giving birth and the presence and number of implantation scars in their uteri were recorded. Each scar indicates the site of placental/foetal attachment during pregnancy. Scars are roughly 2 mm in diameter and can be counted easily under a microscope (Bacon and McClintock, 1994). The number of pups carried in utero were compared to the number of pups found at birth and thus the amount of post-implantation pup loss could be quantified. The number of implantation scars was compared with number of corpora lutea to detect pre-implantation loss.

Collection of Urine

Urine from domestic cats (*Felis catus*) was used as a source of predator chemical cues. These cats normally hunt for mice and have mice as part of their diet. If needed, additional meat was added to their diet. Freshly voided urine was frozen (-22°C. Once defrosted, urine was used only once. Non-predator urine was obtained from guinea pigs and Norway rats. For collection of the urine from Norway rats held in overcrowded conditions, 10-12 animals were placed in a standard size cage for 14 days. Animals received a reduced per capita ration so as to maintain 80% of normal body mass and water *ad libitum*. Individuals of these species (as appropriate for the type of treatment specified in the experiment) were placed into metabolic stainless-steel cages overnight, and urine was collected and stored using the method described above. Urine was collected and stored at -22°C.

"Open Field" Test with Added Stress (Rough Handling Conditions).

An "open arena" (1.5 x 1.5 m) with bright lights was used. Pregnant females were placed for 15 minutes in the centre of the arena on 1-st, 3-d , 5-th and 7-th day of gestation. During the test, we also used a buzzer, which made a loud noise, every 5 minutes. In addition, rats were handled roughly to physically induce stress. Blood samples were drawn after each test for progesterone and corticosterone assay.

Assay for Progesterone and Corticosterone

Rats within each treatment were randomly assigned to one of four cohorts. Blood samples (300 μl) were obtained from the rear foot pad (Miller et al., 1997) every second day for each cohort for each of the treatment for the first seven days of gestation. This minimizes the handling and sampling of individual rats, while allowing a detailed study of changes in hormonal pattern as a function of time and treatment. Our experience shows that this method of repeated blood sampling has no long-term effect on visible scarring associated with traditional tail sampling technologies (Miller et al., 1997). Samples were centrifuged and the plasma frozen at -20°C until subsequent analysis. Plasma progesterone and corticosterone were assayed (in triplicate) by the coat-a-tube radio immuno assay (RIA) method (Diagnostic Products, Los Angeles, California, ≈50μl/assay; Miller et al., 1997).

Experimental Design

The experimental method consisted of applying 1 ml of a test solution to the bedding of pregnant rats every other day for different time durations. This application maximised the likelihood of physical and odour exposure of the test stimulus to the female. In the first experiment, six treatment levels were used:

1. tap water (WAT), as a negative control;
2. urine from guinea pigs maintained on a vegetarian diet (vegetables, grains and water *ad libitum*), as a urine control (GPU);
3. urine from domestic cats maintained on a feral mouse diet (CU), as a model stimulus representing unadultered predator urine. Cats were maintained on the feral mouse diet for 14 days before urine collection;
4. urine from the same feral cats as CU treated with 4% $HgSO_4$ to yield urine where small peptides, amines, and sulfur compounds were precipitated (CU-$HgSO_4$). The supernatant was used as the test stimulus and did not contain indicators of a carnivorous diet (Nolte et al. 1994);
5. urine from cats on a vegetarian diet (CU-v). After collecting urine from the cats used in the above treatments, we switched the animals to a vegetarian diet (oatmeal, peaches, milk and bread *ad libitum*) for 14 days before urine collection; and
6. urine from overcrowded rats (OWU), as source of conspecific chemical signals of high population density (intrinsic factor for regulation of population density).

After mating , females were randomly assigned to treatment groups: WAT (n=20), GPU (n=20), CU (n=20), CU-$HgSO_4$ (n=21), CU-v (n=26) and OWU (n=21).

Mean differences among treatment groups were determined in separate analyses for the number of pups and sex ratios using a fixed-effects analysis of variance. Post-hoc differences among treatment groups were determined using a Spjotovoll-Stoline test for unequal sample sizes (Statistica; StatSoft Inc, Tulsa, Oklahoma).

RESULTS

The number of pups born to female rats varied as a function of treatment ($F = 28.61$, df = 4,102, $P < 0.001$). Post-hoc analysis indicated that diet, specifically urine products derived from meat diets, was important in producing the reproductive inhibitory effect. Females exposed to urine derived from cats maintained on a mouse diet had the smallest litter sizes (Table 1). All other treatment resulted in litter sizes similar to those seen for females exposed to water (i.e. negative control), except for the group treated with urine from overcrowded rats. For this treatment, litter size was significantly reduced relative to both of the controls. Sex ratios also were affected by the treatment ($F = 16.53$, df = 4,102, $P < 0.001$). Fewer female pups were born to rats exposed to urine derived from cats maintained on mouse diets and to rats exposed to urine from overcrowded conspecifics. All other treatments resulted in sex ratios similar to the negative control, which itself was similar to 0.5.

Table 1. The influence of exposure to predator odours and conspecific odours on alternate days during gestation on litter size and sex ratio in Norway rats

Type of treatment	Litter size	Sex ratio (females/total number of pups)
Exposure to cat urine (CU)	4.8 ± 0.4	0.31 ± 0.07
Exposure to CU treated with $HgSO_4$ to remove indicators of carnivorous diet (CU-$HgSO_4$)	9.3 ± 0.4	0.50 ± 0.11
Exposure to urine from cats on vegetarian diet (CU-v)	8.3 ± 0.4	0.50 ± 0.09
Exposure to tap water (WAT)	8.6 ± 0.2	0.51 ± 0.11
Exposure to guinea pig urine (GPU)	8.6 ± 0.2	0.52 ± 0.12
Exposure to urine of overcrowded conspecifics	6.7 ± 1.2	0.41 ± 0.07

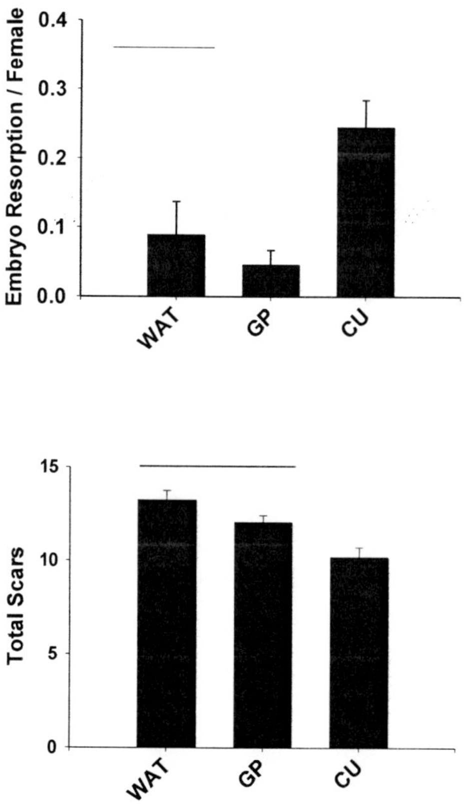

Figure 1 a,b. The influence of exposure to urine odours from cats (CU), guinea pigs (GPU) and tap water control (WAT) on (upper) embryo resorption and (lower) implantation in Norway rats. There were no significant differences between the water and guinea pig urine groups.

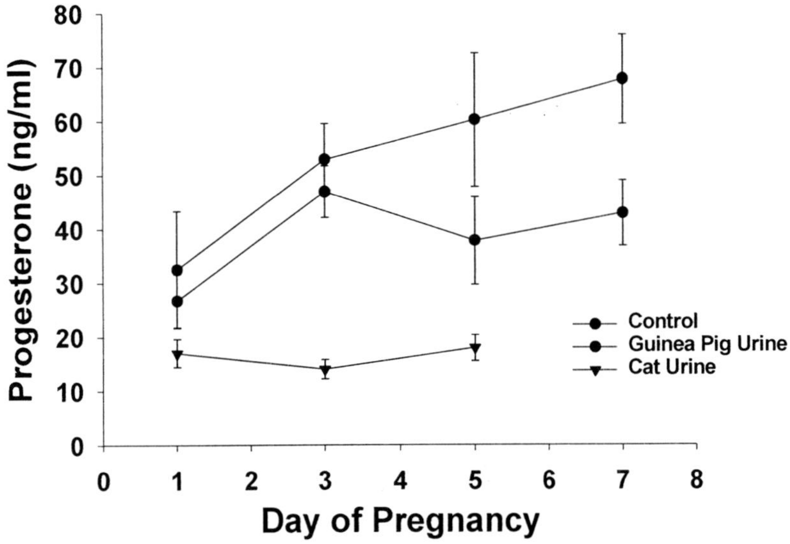

Figure 2. Plasma progesterone levels (mean ± sem) in Norway rats exposed to different urine odours (tap water control, guinea pig urine or cat urine) every other day during gestation.

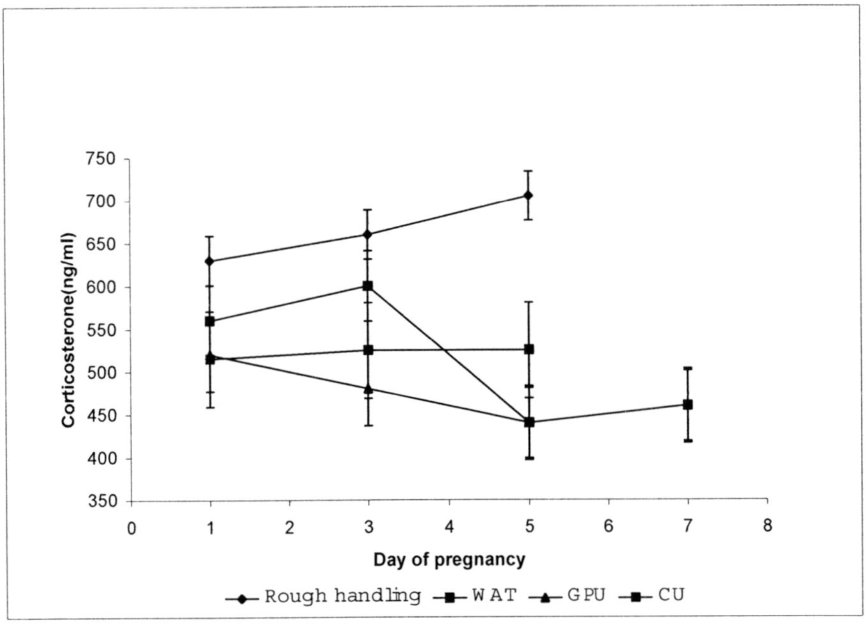

Figure 3. Plasma corticosterone levels (mean ± sem) in Norway rats exposed to urine odours (cat urine, CU; guinea pig urine, GPU; or control tap water, WAT) every other day during gestation or to rough handling in an "open field" test with added stress.

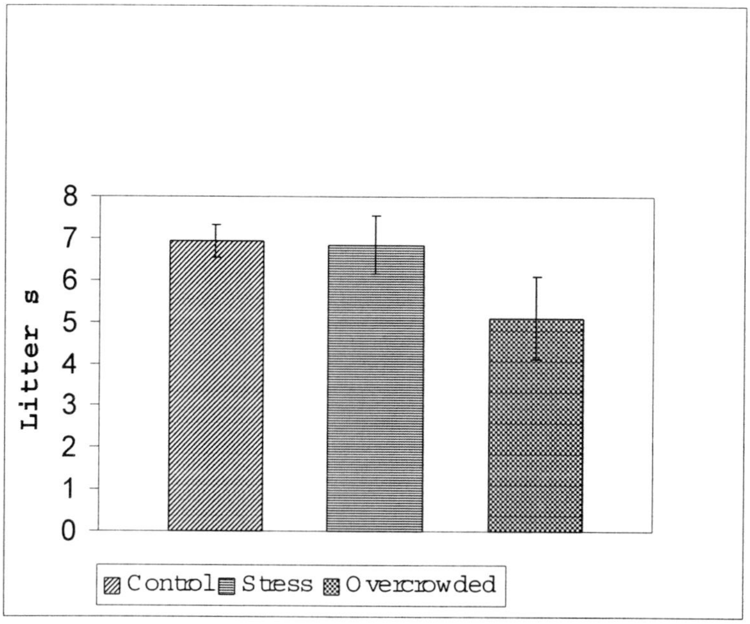

Figure 4. The influence of stress or exposure to urine derived from overcrowded conspecifics on litter size in Norway rats, where : control = exposure to tap water every other day during gestation; stress = "open field" test with added stress; overcrowded = exposure to urine from overcrowded conspecifics every other day during gestation.

Table 2. The influence of exposure to active compounds from predator urine on reproduction in Norway rats. Litter size was significantly reduced ($P < 0.001$), pre- and post-implantation loss was also significantly higher ($P < 0.001$) under active compounds exposure.

Type of treatment	Litter size	Total loss (pre- and post-implantation) %	Number of corpora lutea.
Active compounds (3%)	6,0 ± 4,04	51,41 %	12,3 ± 1,74
Control (WAT)	9,5 ± 3,14	25,40 %	12,2 ± 2,14

When urine was deficient in compounds characteristic of meat digestion, it had no effect on rat reproduction, irrespective of species, or how the deficiency of these products was obtained, i.e. via diet restriction or chemical removal. Urine derived from rats housed in crowded conditions suppressed the reproductive output of rats at rates comparable to rats that were exposed to predator urine. These data suggest that catabolism of muscle mass due to starvation results in urine that contains reproductively inhibitory materials.

The observed changes to litter size, sex ratio, and gestation period brought about by early exposure to predator urine may involve one or more of the following mechanisms: (1) decreased ovulation; (2) delayed, reduced and/or differential fertilisation of the egg by sperm; (3) reduced implantation; (4) differential resorption of fertilised egg or blastocoels; and (5) delayed implantation coupled with sex-linked differential survival of the blastocoel.

Progesterone is a key ovarian hormone produced by the corpora lutea, and is responsible for the maintenance of the fertilised egg, preparation of the endometrium, and maintenance of pregnancy. Factors that disrupt progesterone production could lead to the altered reproductive success observed for rats exposed to predator urine. In follow-up experiments, we monitored plasma progesterone and corticosterone (a hormone released in response to stress) in female Norway rats during early gestation. In addition to groups exposed to urine odours, and control treatment, there was another group in which rats were handled roughly to physically induce stress ("open field" test with added stress). As we observed in our previous studies, female rats exposed to cat urine had smaller litter sizes. Based on the number of corpora lutea and implantation scars, it appeared that reduction in litter size was due to reduced implantation and resorption of embryos during the early part of gestation (Figure 1). Both of the control groups had similar embryo resorption rates, while the urine group experienced a higher embryo resorption rate per female (Figure 1a). These effects were also observed between treatments for the number of implantation scars (Figure 1b). Consistent with the morphological evidence was the observation that plasma progesterone levels were dramatically suppressed in rats exposed to cat urine relative to levels observed in the water control group and for rats exposed to guinea pig urine (Figure 2). We did not observe statistically significant differences in plasma corticosterone levels for rats exposed to predator and non-predator urine, while rough handling of animals caused a clear elevation of corticosterone (Figure 3). Plasma corticosterone levels were unrelated to the two measures of reproductive output, i.e. number of live births and resorption of embryos, with stress per se not influencing reproductive output (Figure 4).

Suppression of rodent reproduction also occured when rodents were exposed to urine of conspecifics housed under high population densities. For urine from overcrowded conspecifics we observed the very same effect as for predator urine (Figure 4). The fact that rodents respond to certain chemical signals in predator urine in a similar fashion may be fortuitous , and may have more to do with the coincidence that the urines contain similar chemical cues resulting from protein digestion in carnivores and protein catabolism in nutritionally deprived rodents rather than specific predator-prey adaptations.

We tested a number of active compounds from predator urine and urine from overcrowded conspecifics (Rattus norvegicus) in Norway rats under laboratory conditions. A mixture of active compounds significantly affected reproduction in Norway rats (Table 2).

Under semi-natural conditions mixture of active compounds (3% encapsulated into food bait) significantly reduced number of reproducing females in House mouse (Mus musculus). In control group percent of reproducing females was 60 %, while in experimental group – 10% only. Application of nest material soaked with active compounds was less effective.

Discussion and Conclusions

Reproductive traits in rodents are affected by a number of environmental, social and chemosensory factors, e.g. the nutritional status of females will influence ovulation rate and litter size (Hamilton and Bronson 1985), as will exposure of females to other rodents of various social status (Steiner et al. 1983; Huck et al. 1988). Other well-described influences include synchronisation of ovulation amongst female cohorts (Whitten 1956), acceleration or delay of puberty (Vandenbergh 1969; Lombardi and Vandenberg 1977), pregnancy block owing to stress, and failure to implant blastocysts when female rodents are exposed to the odour or urine of strange males (Bruce 1959). Frequently the magnitude of these effects is species-and strain-specific (MacNiven et al, 1992).

The majority of these studies on reproductive inhibition have focused on intraspecific influences of semiochemicals and how they influence reproductive output and behaviour in females. A few studies have focused on between-strain influences or interspecific influence, although the source odour generally is still confined to rodents.

During our investigation on the effects of predator odour on rodent reproduction and repellency, we found that female rats exposed to cat urine during pregnancy had reduced litter sizes at parturition (Voznessenskaya 2003). Exposure to predator odour also caused disruptions of the oestrous cycle (Voznessenskaya et al. 1992). These effects bear striking similarities to the studies of the effects of rodent urine odour on intraspecific rodent reproduction. If such similarities are broadly based, then similarities in mechanisms of perception, reproductive physiology, and chemical nature of stimulus might be anticipated.

We do not believe that reduction in litter size is attributable to an adaptive response by rodents to predator odours. Rather, we propose the following interpretation. Urine contains information about the identity of individuals, reproductive status, and dominance status. We postulate that urine also contains information about environmental quality as reflected by nutritional status. Investigation of urine from a variety of sources would serve as an efficient way to integrate environmental information. During times of food depletion, an individual could assess the nutritional status of the population. If food becomes limiting, rodents will begin to catabolise their own muscle protein and the urine will contain larger amounts of protein degradation products. These signals could serve to trigger mechanisms that would affect reproduction. Given that the generation time of rodents is short, complete reproductive inhibition may not be adaptive. However, reduced reproduction may be beneficial. Reduced reproduction would relieve energetic constraints on lactating females that might otherwise jeopardize survival if a full litter size were attempted.

Litters are biased toward producing males when predator or rat catabolic urine is used as a stimulus. This is consistent with theory on reproductive value. Even with reduced litter size, females may still experience lower survival probabilities during reproduction and lactation in food limiting environments because of energetic constraints. However, males would be less constrained by such energetic considerations. Thus, their survivorship probabilities may be higher than females, and by implication their value in contributing to fitness would also be higher. So then, why should rodents reduce reproduction when presented with predator urine ? Predators on rodent diets would produce urine with many of the same rodent-derived metabolic products. It is only coincident that the two urines produce the same effect.

The proposed method utilizes naturally derived compounds that pose no environmental hazard. In nature, predators are one of the most powerful extrinsic factors affecting prey population cycles (Hentonnen et al. 1987; Klemola et al. 1997). At the same time, high population density in rodents is the most powerful intrinsic factor for regulation of population density. Our method utilizes combination of intrinsic and extrinsic factors regulating population density under natural conditions. One of the most serious advantages of this method is lack of habituation to repeated exposures to such types of compounds. At the time we keep a seventh generation of rats in our laboratory under persistent exposures to predator odours. These animals still responding to predator urine exposures with reduced litter size.

Thus, this method should prove useful in reducing our reliance on pesticides with less favorable environmental properties while achieving the goal of reducing rodent populations.

ACKNOWLEDGMENTS

This research was supported by grants from Russian Foundation for Basic Research #98-04-48865 and 01-04-48411, Civilian Research and Development Foundation CRDF # RB1-2036, Russian Academy of Sciences, "Biological Resources" # 3.1.1., Foreign Agricultural Service, USA, #RS72

REFERENCES

1. S.J. Bacon, M.K. McClintock:. Multiple Factors Determine the Sex Ratio of Postpartum-Conceived Norway Rat Litters. *Physiol. Behav.* , **56**, 359-366 (1994).
2. H.M. Bruce: *An Exteroceptive Block to Pregnancy in the Mouse.* Nature (London), 61, 157-195, (1959).
3. G.D. Hamilton, F.H. Bronson: Food Restriction and Reproductive Development: Male and Female Mice and Male Rats. *Amer. J. Physiol.,* **250**, 370-376, (1985).
4. H. Hentonnen, T. Oksanen, A. Jortikka, V. Haukisalmi: How Much Do Weasels Shape Microtine Cycles in Northern Fennoscandian Taiga ? *Oikos,* **50**, 353-365, (1987).
5. U.W. Huck, N.C. Pratt, J.B. Labov, R.D. Lisk:. Effects of Age and Parity on Litter Size and Offspring Sex Ration in Golden Hamsters (Mesocricetus auratus). *J. Reprod. Physiol.,* **83**, 209-214, (1988).
6. T. Klemola, M. Koivula, E. Korpimaki, K. Norrdahl : Small Mustelid Predation Shows Population Growth of Microtus Voles: a Predator Reduction Experiment. *Journal of Animal Ecology,* **66**, 607-614, (1997).
7. J.G. Lombardi, J.G. Vandenbergh:. Pheromonally-Induced Sexual Maturation in Females: Regulation by the Social Environment of the Male. *Science,* **196**, 545-546. (1977).
8. E. MacNiven, D. deCatanzaro, E.V. Younglai:. Chronic Stress Increases Oestrogen and Other Steroids Around Intrauterine Implantation in Inseminated Rats. *Physiology and Behaviour.,* **52**, 159-162, (1992).
9. L.A. Miller, B.E. Johns, D.J. Elias, K.A. Crane: Comparative Efficacy of Two Immunocontraceptive Vaccines. *Vaccine,* **15**, 1858-1862, (1997).
10. *National Economic Statistical Service.* Noncitrus Crop Report. Washington, DC, (1996).
11. D.L.Nolte, J.R. Mason, G. Epple, E.V. Aronov, D.L. Campbell: Why Are Predator Urines Aversive to Prey ? *J. Chem. Ecol.,* **20**, 1505-1516, (1994).
12. *Statistica.* StatSoft, Inc. Tulsa, Oklahoma, (1999).
13. J.G. Vandenbergh: Male Odor Accelerates Female Sexual Maturation in Mice. *J. Endocrinol.,* **84**, 658-660, (1969).
14. V.V. Voznessenskaya, C.J. Wysocki, E.P. Zinkevich: Regulation of Rat Estrous Cycle by Predator Odors: Role of the Vomeronasal Organ. In: Doty, R.L., Muller-Schwarze, D. (Eds): *Chemical Signals in Vertebrates 6.* Plenum Press, New York, 281-284, (1992).

15. V.V. Voznessenskaya, S.V. Naidenko, N.Yu. Feoktistova, L.A. Miller, L. Clark: Hormonal Mechanisms of Litter Reductions in Rodents Under Predator Odor Influence. *Chemical Senses,* **25**, 604-605, (2000).
16. V.V. Voznessenskaya: Predator Urine as a Reproductive Inhibitor of Rats. *Journal of Chemical Ecology,* (2003, in press).
17. V.V. Voznessenskaya, S.V. Naidenko, N.Yu. Feoktistova, G.J. Krivomazov, L.A. Miller, L. Clark: Predator Odours as Reproductive Inhibitors for Norway Rats. In: G.R. Singleton, L.A. Hinds, C.J. Krebs, D.M. Spratt (Eds): *Rats, Mice and People: Rodent Biology and Management*, ACIAR, Canberra, 131-136, (2003).
18. W.K. Whitten: Modifications of the Oestrous Cycle of the Mouse by External Stimuli Associate With the Male. *J.Endocrinol.,* **13**, 399-404, (1956).

Chapter 10

ADDITIONAL PURIFICATION OF WATER BODIES AND WASTE WATERS WITH CONSORCIUMS OF AQUATIC PLANTS CONTAINING MICROORGANISMS

I.N. Gogotov
Institute of Basic Biological Problems RAS, Pushchino, 142290, Russia

ABSTRACT

Phototrophic and chemotrophic microorganisms and their consortia with water plants (aquatic fern *Azolla*, water hyacinth *Eichonia cassipens*) are capable of accumulating metal ions of Ni, Pt, Ru, Cu, Cr, Pb, Zn, Si, Au and other pollutants. This ability enables using them for purification of agricultural and industrial waste water from toxic heavy metals, hydrocarbon and to obtain rare trace metals. For this purpose it is possible to use growing cultures and immobilized cells or enzyme hydrogenase. Purple bacteria (*Rhodobacter* spp., *Rhodopseudomonas* spp.) are able to accumulate Cu, Zn, Ni and Hg., showing various resistance to these metals. Green algae *Chlorella* spp. and water fern *Azolla*, duckweed Lemna spp. and water hyacinth showed higher ability for biosorption of metal ions.

Key words: Aquatic plants, phototrophic microorganisms, hydrogenase, heavy metals, hydrocarbon, purification.

Many metal ions play an important role in a metabolism of microorganisms, plants, animals and man [1,2]. They are required for synthesis of pigments, electron carriers, enzymes, vitamins and other cell components. Elevated concentrations or deficit of a number of metal ions induce a toxic effect on organisms, causing various diseases or suppressing their growth. Chemotrophic and phototrophic microorganisms, as well as their consortia with plants and lichens, accumulate efficiently the metals from an environment in the amount higher than their requirements for growth, that can be used for bioremediation of waste water and soils, and for the development of technologies of obtaining dissipated expensive metals and homeopathic drugs as well [1-4]. The amount of metals accumulated by them can be different. The absorption mechanisms also vary from simple physicochemical process of absorption and desorption, up to highly specific systems of a transport of metal ions in the cells. More than 16 metals are necessary for the growth of microorganisms and plants and their metabolism. However, biological functions of many metals and mechanism of their operation are not clear. The purpose of this presentation is the analysis of the literary and own data on efficiency of a biosorption and stability to ions of metals of the growing cultures, suspensions of native (immobilized) cells and purification of waste water from different pollutants.

1. Biosorption of Metals by Phototrophic Microorganisms

1.1. Purple Bacteria

The purple not sulfur bacterium *Rhodobacter capsulatus* and different species of *Rhodopseudomonas* spp absorb ions of Cu, Ni and Cr from waste water up to 80 – 90 % [1]. Purple sulfur bacterium *Ectothiorhodospira shaposhnikovii*, making a biopolymer [6], is able to absorb more then 90 % Cu, Ni and Zn at a contents 100 mg/l of metal ions in the medium (Table 1). The metal-saturated biopolymer in it reached 25 – 30 % of a dry mass. According to our data, in the presence of 214 g/l Ni^{2+} and 0,00371 g/l Cu^{2+} their absorption was much higher by a biopolymer (60 – 70 %), than by intact cells (30 – 40 %) at pH 6-7.

Metals, absorbed by the cells of purple bacteria or their biopolymers, can be isolated from them, at low pH (~ 2). The reduction of ions metals up to a metal state is possible in the presence of hydrogenase isolated from purple sulfur bacterium *Thiocapsa rosopersicina*, if the intermediate low potential carrier of electrons (ferredoxin, cytochrom C_3 or methylviologene) will be added in a the reaction mixture. Hydrogenase or microorganisms, containing it, reduced ions of Ni^{2+}, Pb^{2+}, Pd^{2+} and Ru^{3+} in these conditions. However, reduction of Cd^{2+}, Co^{2+} and Cu^{2+} ions did not take place, though there are no thermodynamic limitations for this reaction. The rate of metal ions reduction, measured by consumption of H_2, was low (~ 100 nmol/hour) due to the toxic effect of metal ions on the hydrogenase. Investigation of stability of the hydrogenase from *T. roseopersicina* to metal ions has shown, that it considerably depends on temperature and time of an incubation [8,9]. Inhibition of this enzyme in the presence of Cd^{2+} (K_i = 0,95) and Ni^{2+} (K_i = 3,8 mM) depends on pH too. In the presence of Ni^{2+} and Cd^{2+} ions inactivation of hydrogenase is reversible, whereas Cu^{2+} and Hg^{2+} ions induced an irreversible inhibition [1,8.9].

Tabl. 1. Accumulation and stability to heavy metals of prokaryotic phototropic bacteria

Microorganisms	Metals	Accumulations, %/ Stability, μM
PURPLE BACTERIA		
Rhodobacter capsulatus	**Cu, Ni, Cr**	70-80 %
Rhodopseudomonas spp.	Hg, Te, Se	40-50 %
Rhodospirillum rubrum	Hg	35 %
E. shaposhnikovii		
cells	Cu, Ni, Zn	40-45 %
biopolimer	Cu, Ni, Zn	61-99 %
T. roseopersicina	Ni, Cd, Cu	43-48 %
Rh. capsulatus	Hg	0.25 μM
R. rubrum	Hg	0.1 μM
CYANOBACTERIA		
Cyanidium caldarum	Cu Fe Zn Pb	72.0 mg/g d. w. 68.0 mg/g d. w. 3.6 mg/g d. w. 0.4 mg/g d. w.
Synechococcus sp.	Mn, Co, Zn, Ag, Sn, Cs, Hg	55-80 %
Phormidium valderianum	Cd, Co	65-70 %
Nostoc muscorum	Cd, Ti	70-80 %
Anacystis nidulans	Cd	50 μM

A great number of species of the family *Rhodospirillaceae* are investigated with respect of the stability to oxyanions of heavy metals [3]. Species of this family maintain redox-control due to reduction of tellurite (TeO_3^{2-}). For *Rh. sphaeroides* the mechanism of resistance to this group of tellurite involves membrane – bound FAD-dependent metal oxyanion reductase (MOR). As a result the intracellular precipitation of metal and significant hydrogen evolution under anaerobic photoheterotrophic conditions occur. The salts of TeO_3^{2-}

and SeO_3^{2-} (~ 1 μg/ml), added to the cultural medium, kept it from the growth of Gram-negative bacteria.

1.2. Cyanobacteria

Many cyanobacteria have transport systems of metal ions from the environment in the cells. A transport system of Ni^{2+} with K_m = 117 nM, enabling to concentrate it approximately by 2700 times is known for *Anabaena cylindrica*. Absorption of Ni^{2+} depends on membrane potential and decreases in the dark or in the presence of metabolism inhibitors [1]. Cells of *Anacystis nidulans* absorb actively Cd^{2+}. This absorption is inhibited completely by Ca^{2+} and Zn^{2+}. Low concentrations $HgCl_2$ (6 μm) influence on the energy transfer inside phycobilisomes in *Spirulina platensis* and its increased concentrations (18 μm) suppress electron transport from water to methylviologene. The absorption of metals in the cells of *Cyanidium caldarum*, grown in the light, depends on E_h: (mg/g dry weight): Cu (72, E_h 40 mV) > Fe (68, E_h 400 mV) > Zn (3,56, E_h 280 mV) > Cr (1,95, E_h 40 mV) > Pb (0,43, E_h 40 mV).

Passive accumulation of Mn, Co, Zn, Ag, Sn, Sn, Cs, Hg, Np, Pu and Am was determined for *Synechococcus* sp. With concentration factors from 0 (Cs and Np) up to ~ 10^6 in the row: Pu ~ Hg ~ Sn > Am > Zn > Co > Cs ~ Np [1]. Biomass of *Phormidium valderianum* is efficient for biosorption of toxic metals. It absorbs 65 – 70 % of Cd and Co from solution, containing 25 mM each metal ions. Immobilized into polyvinyl particles cyanobacteria absorbed Cd and Co actively. About 80 % bound metal ions were desorbed with 0,1 N HCl.

1.3. Alga

The cells of *Chlorella* and *Scenedesmus* sp. bind most actively Au^{3+}, Ag^+ and Hg^{2+} at pH 2. Irrespectively, live or dead cells are used for biosorption, it takes place their marked affinity to ions of gold [1,2]. The cells of *C. pyrenoidosa* immobilized on the column with silicagel absorb the gold in reversible way. More than 50 cycles of filling the column with tetrachloraurate solution at pH 1,5 and washing it off with HCl and acidic thiourea were replicated without loss of sorbent efficiency. The experiments performed with Au^+-tetrathionate Na and *Chlorella vulgaris* on the same column showed that both algae could be used both for efficient biosorption of gold and its removing. They may be used for extraction of gold from very diluted solutions, including ocean waters.

Relative value of biosorption to general absorption and absorbing capacity of metal ions varies greatly for different types of algae. For *Ankistrodesmus brauni* and *Ch. vulgaris* the binding of Cd with cell walls satisfies 80 % from general absorption (Table 2). *Ch. vulgaris* is the main absorbing environmental component for other metals, including uranium [1]. For living cells of *Ch. vulgaris* the efficiency of binding metals occurs as follows: $UO_2^{2+} > Cu^{2+} > Zn^{2+} > Cd^{2+} > Ni^{2+} > Sr^{2+}$. For cell walls of *Vaucheria* sp. the order of accumulation was: $Cu^{2+} > Sr^{2+} > Zn^{2+} > Cd^{2+}$[1], and for marine algae the following order was obtained: $Hg^{2+} > Ag^+ > Zn^{2+} > Cd^{2+}$ [2].

Tabl. 2. Biosorption of Metals by Alga

Green alga	Metals	Biosorption, % d. w.
Chlorella vulgaris	Cu	< 1.55
	Au	2.10
	Hg	<0.01
Chlorella sp.	Ur	0.026-0.029
	Mn	<0.08
Chlorella regularis	Mo	0.06
	Ur	0.39
	Cd	0.023
	Mn	0.017
	Pb	0.29
Chlorella fucsa	Hg, Al, Be, Au, Cu, Pb, Ni, Zn	
Chlorella pyrenoidosa	Cd	0.11
	Pb	2.28
Chlamydomonas reinhardtii	Fe	0.5-6.6
Euglena sp.	Al	0.2-1.8
	Ba	0.0012-0.0134
	Zn	0.0049-0.0210
	Mn	0.0036-0.0414
	Ni	0.0006-0.0056
	Cu	0.0011-0.0097
	Pb	0.0031-0.0366
	Ur	0.0005-0.0013
	Th	0.0011-0.0039

Au^{3+}, Au^{+} and other complexes of gold could be efficiently bound by algae biomass; concentration of gold in them could reach 10 % of dry weight. Au^{3+} was shown to reduce up to Au^{+} with further elementary gold in colloid state.

Biosorption of metals can be efficient even in the presence of other cations or protons. Ca^{2+}, Mg^{2+}, Na^{+}, Mn^{2+},Zn^{2+},Co^{2+} and Ni^{2+} , but not K^{2+}, inhibit the absorption of Cd^{2+} by *Ch. vulgaris*; Cd^{2+} hinders binding of Mn^{2+}. In various algae the absorption of two-valent ions of Cd, Cu, Zn and Mn is mainly decreased at low pH. However, it depends on concentration of element. *Ch. vulgaris* was shown to be able to absorb Au^{3+}, Ag^{+} and Hg^{2+} at pH 2. Density of cells can also influence on efficiency of biosorption and it is reduced with increasing biomass concentration.

2. Consorciums of Aquatic Plants Containing Microorganisms

Symbiotic consorciums of aquatic plants containing microorganisms is the object of great attention for protecting the environment from pollutants, obtaining renewable materials and fuel. The aquatic fern *Azolla caroliniana,* the aquatic hyacinth *Eichornia crassipens* and various duckweeds *Lemna spp.* grow at high rates on waste waters purifying them from heavy metals, hydrocarbons, aromatic and organic compounds. Their symbionts (cyanobacteria and purple bacteria) due to the availability of hydrogenase, intensively accumulate metals. Thus reducing their ions (Ni, Pt, Ru, Pd) to a metal state. Both whole microorganism cells and their plant consorciums, as well as purified enzymes in the H_2 atmosphere with intermediate electron carrier being available, are capable of it. Due to the high growth rate (1000 plants in 50 days) the aquatic hyacinth produces during a season 20 – 40 t/ha of dry biomass, which contains in a ton up to 20 – 60 kg NPK and 26 kg of protein, which is essentially rich in aminoacids. Most of the metals (~ 80 %) are accumulated by its roots, while leaves can be used as fodder of cattle, for producing paper and biogas. Duckweed is also able to extensively purify waste waters from organic compounds and heavy metals. Its biomass can also be used as a food additives in fish and cattle breeding. It contains 25 – 45 % protein, essentially rich in aminoacids, 42 % of glucose, valuable pigments and microelements. Azolla is also noted for high growth, producing up to 3 kg/m^2 of aw biomass per month, rich in protein (20 – 25 %), valuable metabolites (phitohormones, vitamins, etc.), and are capable of purifying waste waters from hydrocarbons (4,5 %), heavy metals and toxic algae. Its biomass can be used as a food additive in fish breeding, poultry farming, as well as for enriching soils with organic and other biological active compounds.

Our data showed that aqueous fern *A. caroliniana*, representing symbiotic association with cyanobacterium *Nostoc azollae*, purple (*Rhodospirillum* sp.) and chemotrophic (*Artrobacter* sp.) bacteria, as well as other consortia of aqueous plants with microorganisms (*Eichornia crassipens, Salvinia herzogii*) absorbed actively of metal ions (Table 3, 4).

Table 3. Content of metals (mg/kg) in *Chlorella*, *Azolla* and *Eichornia*.

Cultures	Cu		Zn		Pb		Cd	
Chlorella sp.	2083[1]	254[2]	6771[1]	827[2]	208[1]	25[2]	104[1]	13[2]
A. caroliniana	734	86	1405	163	70	8	7,03	0,82
*A. caroliniana**	1786	124	3075	214	79	5,52	19,84	1,38
E. crassipens, leaves	1216	83	1419	97	54	3,7	13,5	0,92
E. cassipens, stem	1024	102	1706	170	51	5,1	3,4	0,34
E. crassipens, leaves**	2410	151	3615	227	120	7,6	12,05	0,76
E. crassipens, stem**	1612	95	9677	569	323	19	32,3	1,9

• Growth at 1 – 20 ppm; ** Biomass in blooming; 1) Ashes; 2) Dry biomass.

Table 4. Efficiency of Ni^{2+} absorption by various sorbents.

Sorbent	Content, mg/l	pH	Absorption, mg/l
Absorbite	100	6,5 – 7,3	8,1
Azolla caroliniana	1000	6,5	43,4
Eichornia crassipens	2	5,0 – 6,6	11,6
Salvinia herzogii	2	5,0 – 6,6	14,4

The data obtained testify that Azolla absorbed actively metals at their low concentration in the medium (1 – 20 ppm). The absorption of metal ions by Azolla occurred in two stages: active and passive. The metals actively absorbed during the grown of Azolla in ponds or other water reservoirs. Passive absorption takes place when the effluent passes through biomass in filter. In these conditions heavy metals are concentrated in biomass of Azolla as much as 500 – 1000 times within 2 – 7 days of the growth and the content of metals in it reaches 1 % of Cu, Cd, Zn, U, Ag and 0,3 % of Cr^{6+} and Ti. About 40 – 60 % of heavy metals were removed from the water.

Total content of metal in Azolla ashes reached 5 %. Active absorption was carried out at pH 4 – 6 with waste water of electronic and electrolysis industry. At passive absorption the filters made from dried Azolla, bound 3,2 % Ag, 4,2 % Cd, 2,7 % Ni, 2,1 % Zn and 3,9 % Cu, containing up to 3000 ppm metal. Total content of metal in Azolla ashes reached 20. Up to 99 % Cd and Ni was removed from industrial waste water, containing 1383 ppm Cd and 2026 ppm Ni. Maximal amount bound by Azolla metal reached 2,8 %. Up to 74 % Cr^{6+} was removed by Azolla from waste waters, containing it in concentration of 137 ppm. Passive processes of metal binding run at pH 3-9 and they are applied in the case of high concentrated metal wastes in electronic and metallurgic industry. *Chlorella* absorbed most actively Cu, Zn, Pb and Cd, and *A. caroliniana* -nickel (Table 3). The content of metals in plants ashes depended both on the type of plants and its separate parts (leaves, stem) and the stage of plant development.

3. Intracellular Accumulation of Metals and their Transport

Absorption of metals dependent on metabolism is the more weak process than passive absorption. It is suppressed at low temperatures, absence of energy source, by inhibitors of metabolism and uncouples of an electron – transport system [1,2]. Nature and composition of culture medium and physiological state of the cells have an influence on the rate of their absorption ability [10]. For the metals, required for the growth and metabolism of microorganisms, there are transport systems of various specificity for their accumulation from the medium. However, even unessential metals are also absorbed by such systems. So, in some bacteria and algae Cd enters through Mn – absorbing system. In cyanobacterium *Anabaena sphaeroides* Zn^{2+} and Co^{2+} suppress the active transport of K^{+} and Na^{+}, disturbing permeability of the cell membrane. There is a dependence of resistance of this cyanobacterium on the growth phase. Its cells exhibit maximum sensitivity to Co at the initial phase of the growth, and minimum are in the end. It should be mentioned that Co ions are an order more toxic that Zn ions (0,05 mg/l). Green algae *Solenastrum* sp. is more resistant to toxic effect of Zn^{2+} (0,12 mg/l). At present the mechanisms of transport participating in accumulation of heavy metals by microorganisms are not yet understood. The amount of metal accumulated at energy – dependent transport may be much higher than at passive absorption, though there are exceptions. *Ch. vulgaris*, *Ch. pyrenoidosa*, *Spirulina platensis* and *Cyanidium caldarum* bind Ca^{2+} and Mg^{2+} weakly, and this hinders binding of heavy metals ions, that is an important advantage of these algae above ion – exchange resins at their remove from hard and waste water [10].

Energy – dependent absorption of Cr by cyanobacterium *Anabaena doliolum* exhibits two- phase behavior and dependence on concentration of chrome [4]. Unlike other metals which are mainly in cationic forms, Cr presumably exists in the form of oxyanion CrO_4^{2-} which is not blocked by anionic components of bacterial shell. There are only a few data concerning ion chrome transport in algae. The data available testify that green algae retain more chrome, as well as Al and Fe, as compared to brawn and red algae. Epiphytic algae, living on plants, exhibit high affinity to atmospheric pollutants and are capable of accumulating heavy metals from air. The content of Cr and Pb was higher in epiphytic alga *Pleurococcus* sp. in places near the highways.

Cr can enter the cells of yeast through nonspecific anion carriers, permease systems, transferring various anions, like sulfate and phosphate [4]. Some tolerant to chromate mutants of *Neurospora crassa* show strong regenerating properties at sulfate transport. The researches conducted showed that toxicity of CrO^{2-} is due to the specific antagonism to sulfate absorption, whereas the toxicity of Cr^{3+} is the result of antagonism to Fe transport.

As for transport and accumulation of metals and their forms by consortia of aqueous plants with microorganisms no reliable data are available. We (Table 3, 4) and other authors gave only the efficiency of biosorption of a number of metals by some organisms and their organs (roots, leaves, stalks) and showed the possibility of their application for purification of industry and agricultural waste waters [1 – 3, 6]. The data obtained are also important for obtaining medical drugs for treatment of microelemental diseases [7].

REFERENCES

1. I. N. Gogotov, N. A. Zorin, O. A. Zadvorny: Accumulation of metals by phototrophic microorganisms and their extraction. In: *Ecology and soils*. Moscow: POLTEX. **3**, 1999. 238 p. (in Russian).
2. G. M. Gadd Accumulation of metals by microorganisms and algae. In: *Biotechnology* (H.-J.Rehm ed.). VCH: Weinheim. 6 b, 401 (1988).
3. M. D. Moore, S. Kaplan: *ASM News*, **60**, 7 (1994).
4. C. Cervantes, J. Campos-Garcia, S. Devars et al.: *FEMS Microb. Rev*., **5**, 335 (2001).
5. Y. Sag, U. Acikel, Z. Aksu, T. Kutsal: *Process Biochem*., **33**, 273 (1998).
6. H. Schmeichen, H. Witting, S. Martin: *BioEngineering*., №1, 38 (1992).
7. H. Sigel, A. Sigel (Eds): *Metal Ions in Biological Systems. Concepts on Metal Ion Toxicity*. Mir, Moscow, 1993. 366 p. (in Russian).
8. L. T. Srebryakova, N. A. Zorin, I. F. Karpilova, I. N. Gogotov: *Appl. Biochem. Microbiol*., **33,** 282 (1997).
9. O. A. Zadvorny, N. A. Zorin, I. N. Gogotov: *Biochemistry*, **65**, 1525 (2000). (in Russian).
10. W. Ahlf: *Appl. Microbiol. Biotechnol*., **238**, 512 (1988).

Chapter 11

Biosorption Handling of Machine-Building Plant Effluent

R.R. Karimov, D.Y. Morozov, M.V. Shulaev, and V.M. Emelyanov
Kazan State Technological University, 420015 Kazan,

Abstract

The adsorption clearing of model waste water containing lubricant cooling liquids and heavy metals ions as contamination, with use of various adsorbents is studied. The opportunity of biological and biosorption treatment of model waste water from lubricant cooling liquids and heavy metals ions in kinetic and dynamic conditions was also studied.

Key words: lubricant cooling liquids, biosorption, biological clearing, heavy metals ions, waste water, adsorbent, activated carbon, ceolithcontaining rock.

Introduction

Mechanical engineering plants are one of the basic sources of polluting the environment. Slags, and wastes are generated as a result of mechanical and galvanic handling of metals in mechanical engineering plants. The most toxic among them are galvanic sewers, containing heavy metals ions and waste lubricant cooling liquids (LCL). LCL toxity is stipulated by the content of mineral oil, stabilized surfactance and various organic additives in it [1].

Organization of utilization processes of LCL and heavy metals ions has its characteristics and difficulties. At the same time there is experience of their handling in industry including combinations of mechanical, chemical, physicochemical, thermal and biological methods [2,3].

Independent examinations on utilization of these wastes by biosorption method in biosorber of a horizontal type in anaerobic conditions with the use of biocenose artificially obtained on the basis of oxybiotic fissile silts of biological clearing buildings at one of the largest chemical plants were carried out. The drumtype intermixing device with movable granulouse adsorbent is a specific feature of biosorber.

The main principle of biosorption technology is microbiological utilization of various waste water components at the presence of adsorbents. Thus the microorganisms of fissile silt develop on the of adsorbents surface and at the expense of that the efficiency of their operation [4] raises.

EXPERIMENTAL PART

Both pure cultures, as for example bacteria p. Aeromonas and p. Pceudomonas, and microbial communities are used for treatment of wastes [5]. Mixed population of adapted microorganisms prepared on the basis of microbial community of fissile silt biological treatment plant at one of the chemical enterprises was used in this work. Activated charcoal SKT-3, and also ceolithcontaining rock (CCR) - diatomite were used as adsorbent. The process of biosorption removal of highconcentrated organic impurities should take place in anaerobic conditions. In this connection, it is necessary to carry out additions preparation of fissile slit for operation at the definite enterprise.

Fissile silt for further experiments was selected from airtank and was prepared by fermertation. The preparation of biomass for its usage in anaerobic conditions included fermentation of oxybiotic fissile silt at temperature 32- 37 C° with addition of nutritions within several days. In fermentation process the rough eduction of biogas was marked, which composition includes hydrogen sulphide. Biomass instead of brown has gained black colour.

Originally kinetic experiments on extraction of contaminations by various adsorbents were carried out without use of fissile silt originally were carried out. The purpose of these experiments was the clearing up of activity of adsorbents used to various pollutants.

The experiment was carried out as follows: in to 5 flat flasks with were placed capacity 250 mls shots of various adsorbents and added 100 mls of model waste water containing this or that contamination. Dilute LCL and heavy metals ions, namely Cr (IV) and Fe (II) were used as contaminations.

After blowing-down by nitrogen flask was densely closed by fuses and aggressively being shaken on rotation unit during given period of time.

The kinetic curves of adsorption are shown in figures 2-5.

Further experiments on study of kinetics of biosorption clearing were carried out. The course of experiments is similar above to mentioned the difference is only that in flasks there was a mixture consisting of 50 mls of model waste water and 50 mls of anaerobic fissile silt instead of 100 mls of model waste water. Control experiment on biological clearing was parallely carried out and, in this case, adsorbent in flasks was not put.

Then a mixture of an adsorbent and fissile silt was filtered off and the value of COD in filtrates was determined by the accelerated method, and residual concentrations of heavy metals ions was determined by a photocolorimetric method. Knowing terminating concentrations of component at various time of contact of contaminations with fissile silt or mixture of fissile silt and adsorbent we have constructed kinetic curves of concentration change of the relevant contaminations (fig.6-8).

The experimental results in dynamic conditions are obtained on a pilot plant for study of biogetter clearing processes. The plan of a pilot plant is given in fig. 1.

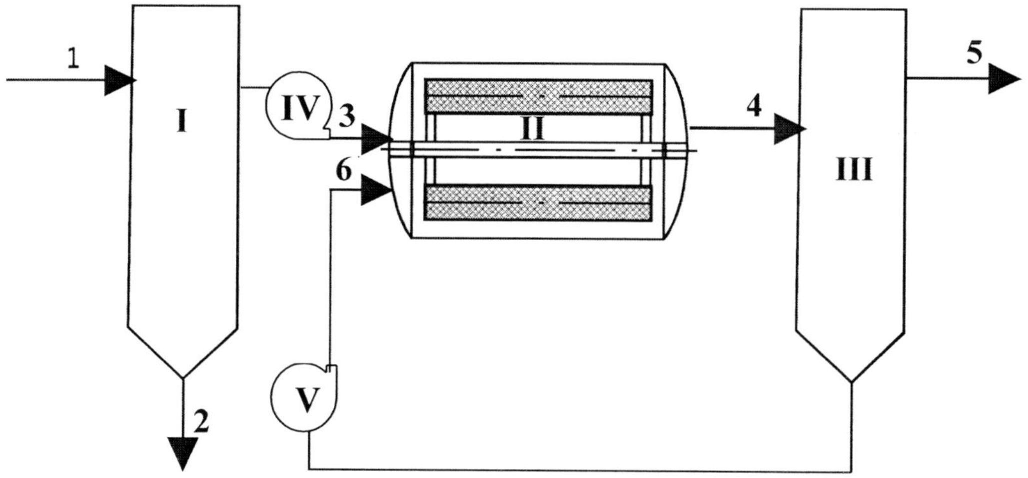

Fig. 1 - Plan of experimental device of biological and biosorption clearing of waste water from heavy metals ions:

I - preliminary sedimentation tank;
II - biosorber of a horizontal type with the drumtype mixing device;
III - secondary sedimentation tank;
IV, V - pump;
1 - model waste water;
2 - mechanical impurities;
3 - sink without mechanical impurities;
4 - handled fluid mixed with fissile silt;
5 - treated water;
6 - recycle of fissile silt.

The model waste water 1 was released from suspended matters in a preliminary sedimentation tank I and was moved with the pump IV on the basic stage of handling - biosorption, or biological clearing (depending on experiment), taking part in apparatus II. From here the mixture of model water and fissile silt 4 was sent to a secondary settling tank III, where the separation of silt from treated water was done. The fissile silt from a secondary settling tank was pumped over back in biosorber II with the help of the pump V.

Horizontal bioreactor with the intermixing drum-type device with a load from a granulouse adsorbent was used. The intermixing device of a frame type was used at biological clearing.

In case of biosorption clearing the particular amount of an adsorbent was filled, depending on contamination at assembling of a reactor in barrels of the mixing device. Then 4,5 l same of the prepared slit suspension was filled in to the assembled bioreactor of capacity and the intermixing was organized. The conditional - pure water was flooded in to a secondary settling tank with capacity 3 l and the recycling of fissile silt was organized with the help of the pump. Then the delivery of model waste water from a preliminary sedimentation tank was organized with the help of the pump. The residence time in the apparatus with the account of recycling was 1-7 day.

RESULTS

Study of a kinetics of adsorption of LCL and heavy metals ions from model waste water (fig. 2-5) has shown, that the granulouse absorbite SCT-3 and diatomite is exhibited satisfactory adsorption properties to studied impurities.

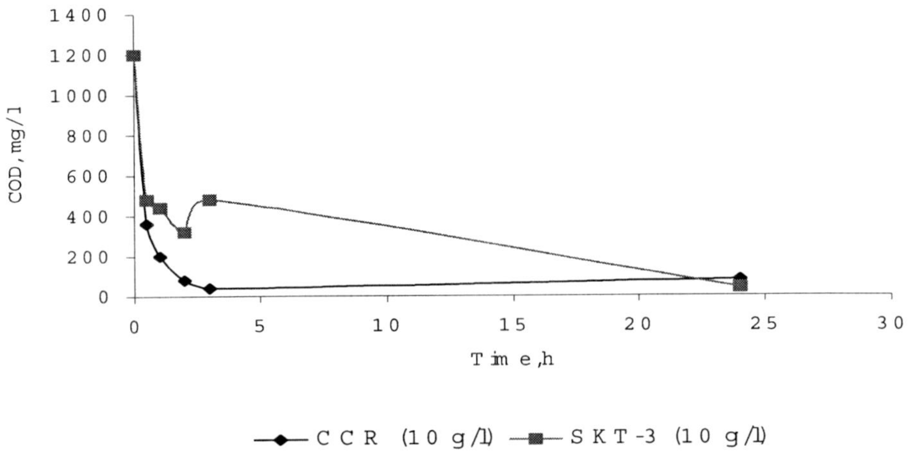

Fig. 2 - Kinetic curves of COD change at adsorption clearing of waste water from LCL

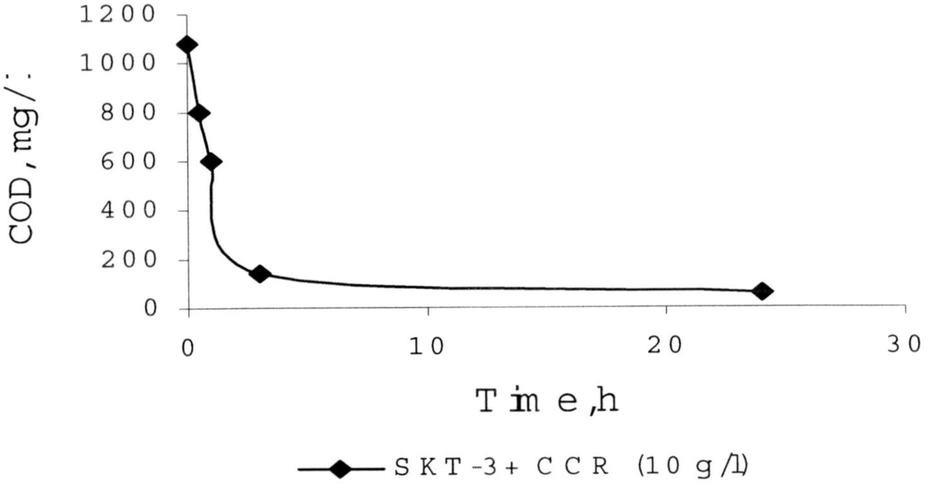

Fig. 3 - Kinetic curves of COD change at adsorption clearing of waste water from LCL

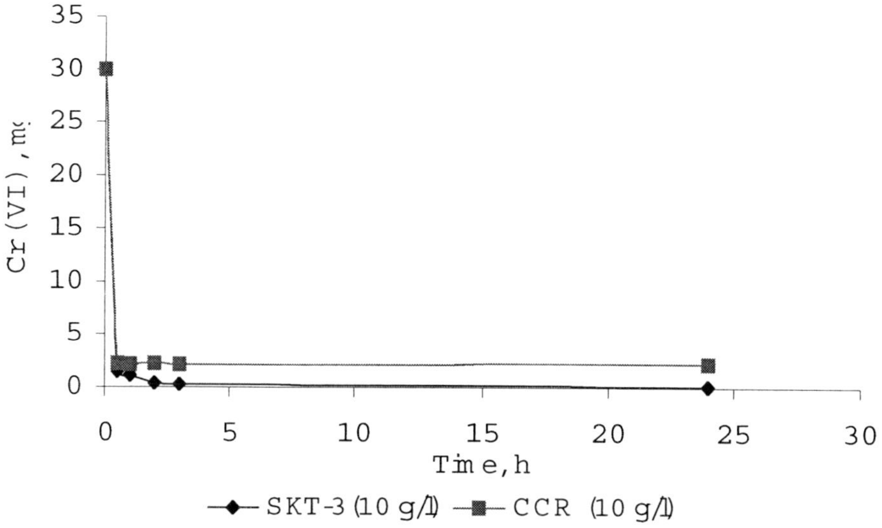

Fig. 4 - Kinetic curves of ion concentration Cr (VI) change in model waste water during adsorption

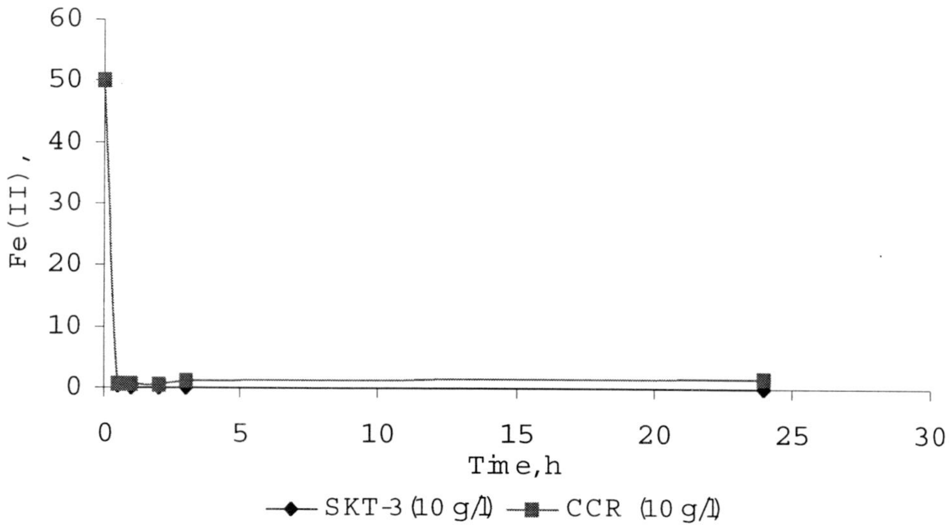

Fig. 5 - Kinetic curves of ion concentration Fe (II) change in model waste water during adsorption

The kinetic curves of biosorption clearing parameters change during given time (fig. 6-8) were constructed according to obtained experimental data.

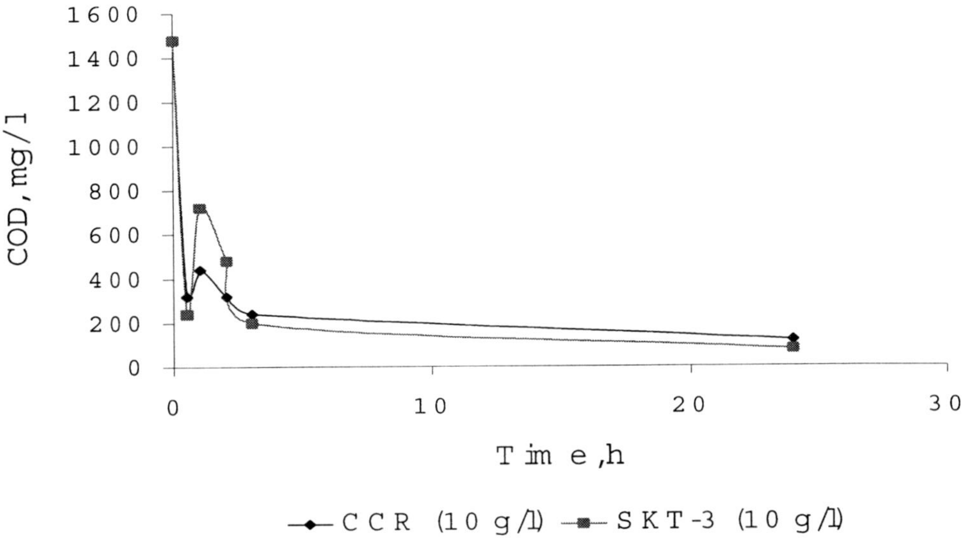

Fig. 6 - Kinetic curves of COD change at biosorption clearing of waste water from LCL

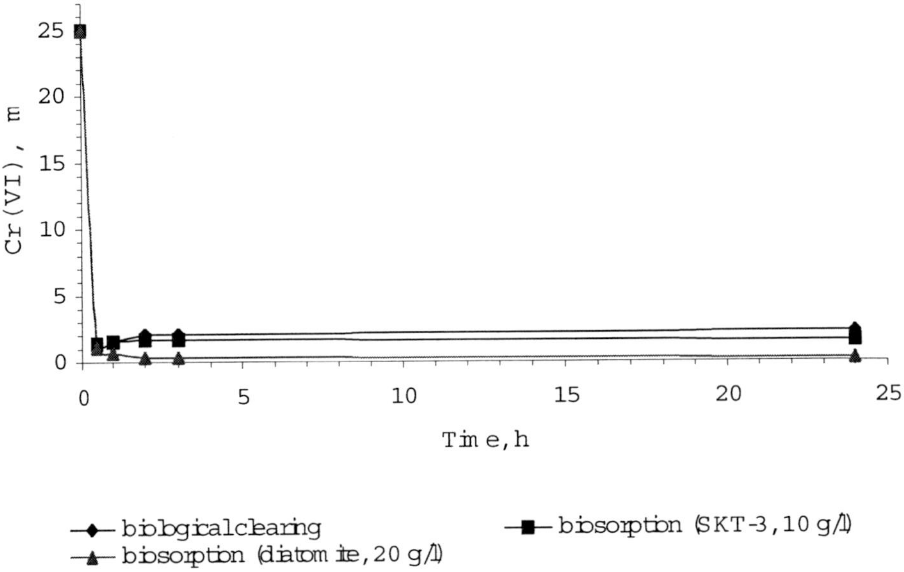

Fig. 7 - Kinetic curves of ion concentration Cr (VI) change in model waste water

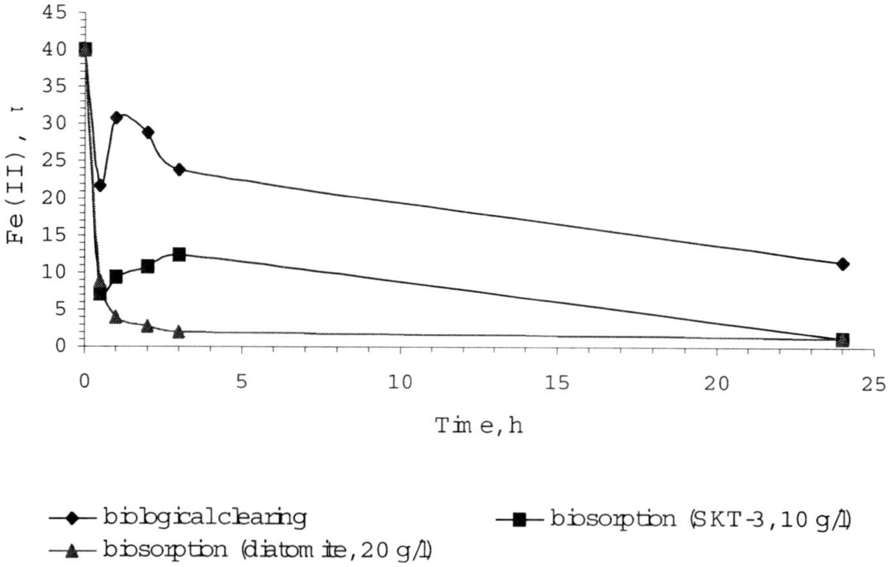

Fig. 8 - Kinetic curves of ion concentration Fe (II) change in model waste water

It is seen from figures 6-8 that the intensive change of polluting substances concentration happens within the first 3 hours of the experiment, but more long time of contact of fissile silt and contaminations is necessary for reaching smaller values of concentrations.

Therefore experiment was carried out in dynamic requirements during 1 and 7 days at clearing from LCL and heavy metals ions, accordingly. The concentration change of pollutants during dynamic experiments is shown in figures 9-10, data on clearing efficiency are given in tables 1-8.

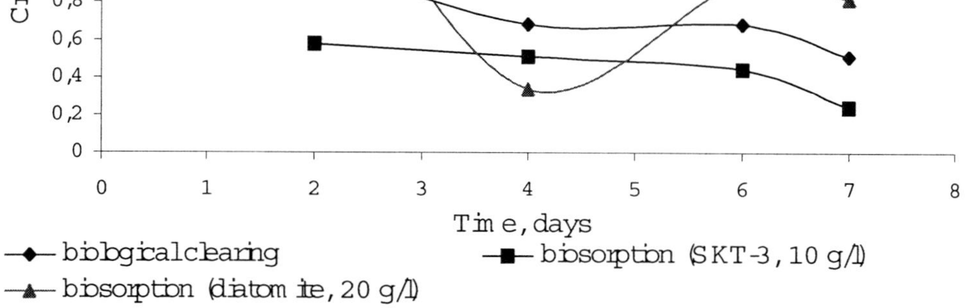

Fig. 9 - Change of ion concentration Cr (VI) in model waste water in dynamic conditions

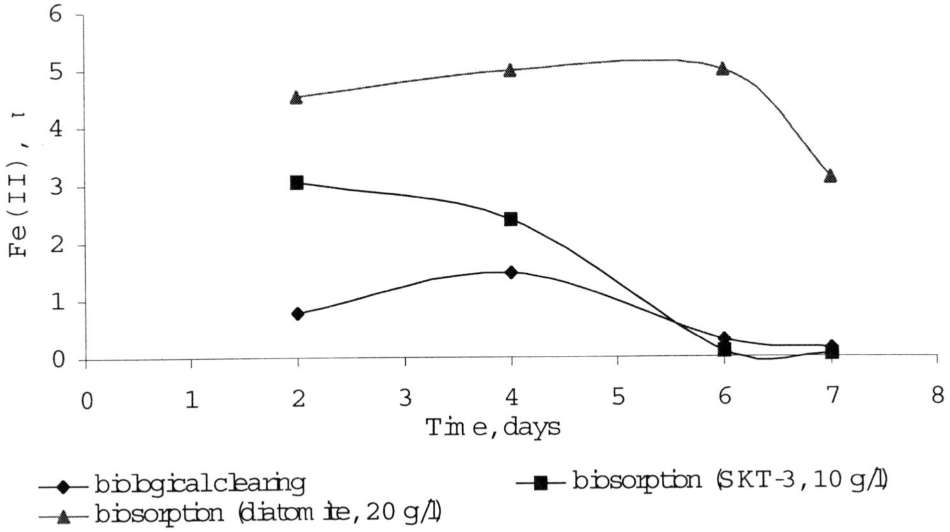

Fig. 10 - Change of ion concentration Fe (II) in model waste water in dynamic conditions

Table 1 - Dynamics biological and biosorption process of clearing LCL

Time,h	Biological treatment				Adsorption (SKT-3)			
	COD, mg/l	pH	T, °C	Degree of purification, %	COD, mg/l	pH	T, °C	Degree of purification, %
0	2300	8,0	20	–	2300	7,9	20	–
24	840	8,25	23	63,5	640	8,3	24	72,2

Table 2 - Dynamics biological and biosorption process of clearing LCL

Time,h	Biological treatment				Adsorption (SKT-3)			
	COD, mg/l	pH	T, °C	Degree of purification, %	COD, mg/l	pH	T, °C	Degree of purification, %
0	3120	8,0	20	–	3120	7,9	20	–
24	680	8,35	23	78.2	600	8,2	24	80.8

Table 3 – Dynamics of biosorption process LCL

Time,h	Adsorption (Diatomite)				Adsorption (SKT-3)			
	COD, mg/l	pH	T, °C	Degree of purification, %	COD, mg/l	pH	T, °C	Degree of purification, %
0	1800	8,0	20	–	1800	8,05	20	–
24	300	8,15	23	83,3	340	8,25	24	81,1

Table 4 – Dynamics of biosorption process LCL

Time,h	Adsorption (Diatomite)				Adsorption (SKT-3)			
	COD, mg/l	pH	T, °C	Degree of purification, %	COD, mg/l	pH	T, °C	Degree of purification, %
0	2200	7,95	20	–	2200	8,0	20	–
24	220	8,35	23	90	480	8,3	24	78.2

Table 5 – Dynamics of biosorption process LCL

Time,h	Adsorption (Diatomite)				Adsorption (SKT-3)			
	COD, mg/l	pH	T, °C	Degree of purification, %	COD, mg/l	pH	T, °C	Degree of purification, %
0	3400	8,05	22	–	3400	8,05	22	–
24	1560	8,2	22	54.12	1520	8,25	21	55.29
48	120	8,25	21	96.47	60	8,25	23	98.23

Table 6 – Biological clearing of waste water from heavy metal ions

Time, day	Cr(VI)			Fe(II)			T, °C	pH
	Initial, Mg/l	Final, mg/l	Degree of purification, %	Initial, mg/l	Final, mg/l	Degree of purification, %		
2	25	1,002	96,0	40	0,772	98,1	25	6,5
4	25	0,678	97,3	40	1,474	96,3	23	6,5
6	25	0,678	97,3	40	0,301	99,2	25	6,5
7	25	0,51	98,0	40	0,15	99,6	26	6,5

Table 7 – Biosorption clearing of waste water from heavy metal ions (SKT-3, 10 g/l)

Time, day	Cr(VI)			Fe(II)			T, °C	pH
	Initial, mg/l	Final, mg/l	Degree of purification, %	Initial, mg/l	Final, mg/l	Degree of purification, %		
2	25	0,578	97,7	40	3,053	92,4	25	6,5
4	25	0,51	98,0	40	2,386	94,0	23	6,5
6	25	0,442	98,2	40	0,1	99,8	25	6,5
7	25	0,238	99,0	40	0,05	99,9	26	6,5

Table 8 – Biosorption clearing of waste water from heavy metal ions (Diatomite, 10 g/l)

Time, day	Cr(VI)			Fe(II)			T, °C	pH
	Initial, mg/l	Final, mg/l	Degree of purification, %	Initial, mg/l	Final, mg/l	Degree of purification, %		
2	25	1,663	93,3	40	4,526	88,7	22	6,5
4	25	0,339	98,6	40	4,982	87,5	22,5	6,5
6	25	0,916	96,3	40	4,982	87,5	23	6,5
7	25	0,815	96,7	40	3,123	92,2	23	6,5

SUMMARY

The adsorption clearing of model waste water containing LCL and heavy metal ions as contamination, with use of diatomite and SKT-3 as adsorbents is studied.

The opportunity of biological and biosorption clearing of model waste water from LCL and heavy metals ions in kinetic and dynamic conditions also studied.

According to experimental data it was found out, that the degree of clearing of waste water by a biosorption method is higher, than at biological clearing.

The efficiency of various adsorbents is studied while using them in the biosorption process. It is shown that the best results on removal of heavy metals ions from model waste water are achieved at biosorption clearing with use of granulouse absorbite SKT-3. However diatomite yields clearing and taking into account its low cost, allows considering diatomite a perspective adsorbent for clearing waste water from LCL and heavy metal ions.

REFERENCES

1. Anserov Y.M., Durnev V.D. Mechanical engineering and protection of an enviroument. *L. Mechanical engineering*, 1989, 224p.
2. Nazarian M.M., Efimov V.T. *Electrocoagulators for clearing industrial sinks*. Kharkov: Your school, 1983, 144p.
3. Nazarian M.M., Shamsha L.F., Schepilov N.S. Study of ozonization process waste LCL of machine-building factories // Information Kharkov Polytech Institute. *Chem.: mechanical engineering*, 1980, issue 10. №17, p. 47-51.
4. *Modern technological concepts of oxybiotic clearing of waste water* / A.S. Sirotkin, S.A. Ponkratova, M.V. Shulaev; Kazan state technological University. Kazan, 2002. – 164p.
5. *Biological clearing of chromecontained industrial waste water* / E.I. Kvasnikova, N.S. Serpokrylov, T.M. Kljushnikova etc.; AS Ukraine. Institute of a microbiology and virusology named after D.K. Zabolotnogo - Kiev: Naukova dumka, 1990. – 112p.

Chapter 12

BIOREMEDIATION OF SOILS DEGRADED AS A CONSEQUENCE OF OIL CONTAMINATION AND SALINATION WITH OIL-FIELD WATER

V.M. Kurashov [1,2], ***T.V. Sakhno*** [1,2], ***A.A. Kolesnik*** [3] ***and V.S. Gavrilov*** [3]
[1]Kazan State Technological University, Kazan, 420015, Russia
[2]Research and Production Centre "Ecological Consortium", Kazan, Russia,
[3]Central Directorate of Natural Resources and Preservation of Environment of the Ministry of Natural Resources of Russia in the Tatarstan Republic, Kazan, 420059, Russia

ABSTRACT

There is a problem of no effect of conventional biopreparations for oil decomposition in the soils of higher oil and salts content, as these higher contents are disastrous to most bacteria species. To solve this problem we used the immobilized compositions, the action of which aimed at both the soil desalination and the oil biodegradation. The immobilized compositions in their different variations have the following content: adsorbent-carrier of inorganic or organic origin; hydrocarbon-oxidizing bacteria; aerobic chemoorganotrophic halophilic archaebacteria. It was found that when desulfurizing the salinited soils, the use of halophilic bacteria in symbiosis with hydrocarbon-oxidizing bacteria protects the hydrocarbon-oxidizing bacteria from destructive effects of the salts. The factor of symbiosis allows the hydrocarbon-oxidizing bacteria to realize their vital functions under extreme conditions (surplus of oil and salts).

Key words: bioremediation, immobilization, oil products, salt affected soil, microorganisms

INTRODUCTION

Oil contaminations as any other organic pollutant, seems to be an additional source of carbon and energy to be injected to the ecosystem. Nowadays several approaches to liquidation of oil-contaminated territories and water areas exist. There are mechanical devices for for gathering oil at water surface, sorbents of different origin, chemical emulators –

detergents, biological preparations on the base of microorganisms oil-destructors. However, not one in mentioned above methods is universal and has no drawbacks. The most perspective trend is likely to be biochemical degradation of oil with the help of microorganisms-destructor [1-3].

It is known that microorganisms determine an intensity of self-purification of water areas and soils from oil products. Hydrocarbondegrading microorganisms are the constant component of microbiocenocys of soil and water areas. However, as a result of accidents, a concentration of oil and oil-products axceed the limiting consentration that leads to inhibition of microbiocenocys. In this connection immobilized microorganisms could play an important role among the steps to increase the productivity of bio cleaning of water areas and territories from oil and o l-products. Adsorption-biological method based on the use of different carriers for microorganisms with well-developed sorption surface becomes very popular nowadays. Immobilization of the cell on the carriers seems to increase the effectiveness of oil contaminations biodegradation [4]. Sorption immobilization method is rather competitive in relation to up-to-date methods based on the mechanical binding of cells [5,6]. At the same time this method is cheap, universal, has no stress influence on cells and is very simple. The use of fixed microorganisms-destructors allows to provide a stability of degrading process, its resistance to various changing in the environment [2]. However, in spite of explicit advantages and perspectives, biosorption method is not widely used in cleaning of natural objects from oil pollutants. Microbial method occupies ~24% of all soil and water areas purification methods; biosorption method – 10% [2].

A purpose of this work is searching of some opportunities for using immobilized biological preparations for water and soil cleaning from hydrocarboncontaining contaminants. Works in this areas supports remediation of nature potential and restore agricultural objects, forests, and bodies of water back into the national economics. The results of the research are used in remediation of soils degraded as a result of both oil contamination and salination with oil-field water on the territory of the Republic of Tatarstan.

Methods and Materials

Experiments for cleaning of oil-contaminated soil were carried out on Vostochno-Suleevskii territory of Romashkinskii deposit of Republic Tatarstan on 2 plots. Both areas have waterproof clay. The cause of the pollution on the first plot (N1) was a break of oil pipeline. The square of pollution is 100%, the square of the pilot area was 1 hectare. Initial oil concentration in soil is 54 g/kg. The content of elements in water extract is placed in the Table. It can be seen from this Table that the salting type of soil is chloric. Initial bacteria content in soil is 1.8 million per 1 g of soil, which is 1.6 times lower the norm. The low bacteria content characterizes the lowering of biological activity of soil. There was no vegetation at the soil surface. Salting with oil field wastewaters and oil contaminants causes defects in fertility of soil. Plot N1 was ploughed 20 cm deep and was treated with 375 kg of immobilized composition (KL) of the following components: adsorbent carrier acidic lignin - 368.5 kg, bacteria of Pseudomonas Putida genus - 3 kg, aerobic heterotrophic halophilic archebacteria - 1 kg, hemolithoautotrophic halophilic bacteria Thiobacillus genus - 2.5 kg.

At the second plot (N2) a contamination was caused by accident escape of gas-liquid mixture at the well of Bavly deposit. Oil contamination in soil is 120 g/kg. Data of analysis of water extract of soil at both plots: chloride-ions 0.45%, sodium ions 0.27%. Exchange sodium 42.07 mg-equivalent per 100g. The square of pollution is 100%, a degree of salting is very high. A bacteria content in soil is 0.3 millions per 1g of soil that 10 times lower a norm. Low bacteria content corresponds to the lowering of biological activity of soil. The content of actinomycetes is 0.1 millions per 1 g of soil (10 times lower the norm).

The plot N2 was plough up to 40 cm and divided into 9 test and 12 pilot micro plots of 4 m^2 each. A surface of the first test micro plots was not treated at all. The surface of other 8 test micro plots was covered with 15 g of hydrocarbon degrading microorganisms "Devoroil" (heteroculture based on Pseudomonas aeroginose, Phodococcus and Candida strains) (D), 15 g of halophylic microorganisms (HPh), 15g of mycelium of NK type (MNK), thermally treated wastes of poultry farm (WPF), osseous powder (OP), swelled vermiculite (SV); D together with HPh, D together with HPh and MNK.

In pilot variants different compositions consisting of mentioned above microorganisms D, HPh, MNK, immobilized at solid adsorbents of WPF, OP, and SV type were used. Compositions were marked as follows:

Halophylic microorhganisms immobilized at wastes of poultry farm HPh-WPF, at osseous powder – HPh-OP, at swelled vermiculite – HPh-SV; heteroculture on the base of Pseudomonas aeruginosa, Phodococcus and Candida (D), immobilized ar wastes of poultry farm D-WPF, at osseous powder D-OP, at swelled vermiculite D-SV.

Hydrocarbondegradin microorganisms D together with halophylic microorganisms HPh, immobilized at wastes of poultry farm, at osseous powder, and at swelled vermiculite were marked as D-HPh-WPF, D-HPh-OP, and D-HPh-SV. Hydrocarbondegradin microorganisms D together with halophylic microorganisms HPh and mycelium MNK, immobilized at EPF, OP, and SV were marked as D-HPh-MNK-WPF, D-HPh-MNK-OP, and D-HPh-MNK-SV.

The reduced bacteria count reflected the lowering of biological activity of soil. There was no plants at the soil surface. The fertility was reduced due to the salting by oil production waste water and oil contamination of soil.

A mineral feeding was performed at a rate of 0.8 g of ammonium phosphate plus KNO_3 per square metre of oil at the first plot at all micro plots of the second plot. The soil was made friable to a depth of oil penetration once a week. Humidity was created by adding fresh water up to the soil's total hydrocapacity. Ten samples (10 g each) were withdrawn from the upper layer of soil after 20 days during 80 days at the plot N1, and during 120 days at the plot N2; they were thoroughly averaged out and were subjected to a study of the biodegradation of oil spills through the residual content of oil in the soil by means of the weight method.

The effectiveness of biodegradation (E) has been obtained by the following formula:

$$E = [(C_0 - C_x) / C_0] \cdot 100\% \quad (1)$$

where C_0 is a concentration of the initial oil quantity in the medium, g/kg; g/l, C_x is a concentration of the current oil quantity in the medium during biodegradation, g/kg; g/l.

RESULTS OF PILOT-INDUSTRIAL TESTS AND DISCUSSIONS

Biodegradation process in soil is poorly known. Difficulties appear, in part, due to the variable composition of soil.

The implementation of the majority of hydrocarbondegrading microorganisms' types for soil remediation is effective when oil concentration is 20-25 g/kg. Higher concentrations lead to inactivation of oxidizing activity of bacteria and then to its death because of inhibiting character of hydrogencarbon substrate. Using of immobilized microorganisms provides an opportunity for applying hydrocarbon-degrading bacteria for oil biodegradation at a concentration of oil exceeding the limiting one.

A character of changing in oil concentration, effectiveness of oil biodegradation, percentage of chloride-ions and potassium during microbiological purification of oil contaminations at plots NN1,2 with the use of immobilized compositions are reflected in tables 1 and 2, respectively.

After 80 days since the treatment the positive changing in soil has been noticed at plot N1. The soil fertility was restored. Chemical and microbiological analyses is presented in the Table 1.

Table 1. Chemical and microbiological analyses

Indicators of oil spills biodegradation and the soil desalting processes over the pilot period after injecting the immobilized composition KL to the soil					**Indicators of biological activity of soil over the pilot period treating the affected soil with immobilized composition KL**	
Time, Days	Oil concentration in soil, g/kg	Water extract analysis,%		Na exchanged, mg-equivalent per 100 g	Numbers of soil bacteria	
		Na^+	CL^-		Bacteria growing on KAA, mln per 1 g of absolutely dry soil, actinomysetas	Bacteria growing on MPA, mln per 1 g of absolutely dry soil
0	54	0,29	0,49	41,29	0,2	1,8
20	46	0,25	0,41	28,84	0,8	3,2
40	31	0,18	0,29	12,05	1,5	5,4
60	8	0,06	0,1	4,49	2,5	8,8
80	0,25	0,03	0,02	2,1	2,7	10,6

The use of immobilized compositions at the plot N1 allows to purify the salt affected soil from oil products with oil concentration exceeding the limiting one for microbiological soil purification (25 g/kg) by a factor of two and to shorten the biodagradation process.

At the test micro plots of the plot N2 during 120 days the oil concentration was not changed (120 g/kg); the results of water extract analysis do not changes as well (chloride ions 0.45%, sodium ions 0.27%), and the quantity of bacteria in oil.

At the pilot microplots of the plot N2 after 120 days a fertility of soil was restored. Oil concentration at the pilot plots was lowered to the limiting concentrations and amounted to 250-750 mg/kg. In water extract one can notice the lowering of chlorine concentration till 0.004%-0.02%, sodium – till 0.005%-0.03%. A quality of bacteria were increased to 10.6-11 millions per 1 g of soil Table 2.

Table 2. An index of oil spills biodegradation-desalting processes during treatment period after injecting of the modified immobilised compositions into the soil

Immobilized compositions, type	Effectiveness of oil biodegradation in soil, %	Days	Residual oil content in soil, g/kg	Water extract analyses, %		Na exchanged, mg-equivalent per 100g
				Na^+	CL^-	
D-WPF	62	120	45,6	0.15	0,25	19,23
D-SV	60	120	48	0,17	0,28	22,5
D-OP	53	120	56,4	0,18	0,29	24,49
HPh-WPF	30	120	84	0,23	0,38	42,07
Hph-SV	27	120	87,6	0,24	0,40	34,12
HPh-OP	22	120	93,6	0,25	0,41	35,11
D-HPh-WPF	99,8	120	0,24	0,002	0,004	0,05
D-Hph-SV	99,5	120	0,6	0,005	0,008	1,09
D-HPh-OP	98	120	2,4	0,03	0,02	2,83
D-HPh-MNK-WPF	99,8	90	0,24	0,002	0,004	0,05
D-Hph-MNK-SV	99,5	90	0,6	0,005	0,008	1,10
D-HPh-MNK-OP	98	90	2,4	0,03	0,02	3,00

Basing on the results of industrial tests at the plot N2 the following conclusion should be made:

The use of immobilized microbiological compositions D-HPh-WPF, D-Hph-SV, D-HPh-OP and D-HPh-MNK-WPF, D-Hph-MNK-SV, D-HPh-MNK-OP allow to clean salted soil from oil (oil products) with oil concentration that 5 times higher than the limiting concentration (258 g/kg) at which a traditional microbiological purification could be made. It allows also to shorten a duration of oil biodegration process.

Thus, as the results of carried out works the following positive processes which determine the restoring of soil fertility took place:

1. The restore of biological activity of soil owing to the injecting of biological preparations and creation the favourable conditions for the vital activity of the available soil bacteria.
2. The use of immobilized compositions allows to purify the salt affected soil from oil products with oil concentration exceeding the limiting one for microbiological soil purification (25 g/kg) by a factor of two and to shorten the biodagradation process.

The results of this work are being applied to the restoring of soil being degraded due to the mutual contamination by oil and salted oil-field wastewater on the territory of the Republic of Tatarstan.

REFERENCES

1. R.I. Mironova, V.P. Noskova, G.E.Rasulova, V.P. Kholodenkov: Biodegradation and biosorption of floating oil by natural micromicetas. *Biotekhnologiya J.,* 7, 44-48 p., (1996).
2. N.I. Pavlenko, V.V. Izzheurova, L.M. Khenkina et al.: Using of microbes-destroyers in cleaning sewage from oil products. *Microbiol. J.*, **53**, 91-95 p., (1991).
3. R.M. Atlas, R. Bartha: Biodegradation of petroleum in seawater at low temperatures. *Can. J. Microbiol*, **18**, 1851 p., (1972).
4. T.V. Sakhno, N.I. Philinova, V.M. Kurashov: *Using Immobilized Microorganisms to Intensify the Removal Hydrocarbon Contaminations from Water and Soil. Bioremediation of Contaminated Soils*, 2nd Edition of Marcel Dekker, New York, Chapter 24, 2000, 373 - 382 p.
5. N.V. Doronina, V.A. Ezhov, YU.A. Trotsenko: Aerobic biodegradation of formaldehyde, methanol and methylamine by immobilized cells. Prikl. *Biokhimiya i mikrobiologiya J.*, **33**, 162-165 p., (1997).
6. M.J. Dempsey, G.M. Black, B. Atkinson: *Proc. Eng. Aspects immobilized cell system*, Rugdy, 246 p., 1988.

Chapter 13

APPLYING KERAMSIT-IMMOBILIZED OIL-OXIDIZING MICROORGANISMS AS WATER CLEANSERS TO HELP REMOVE TRACE PETROLEUM PRODUCTS

***T.P. Pirog* [1,2]*, T.A. Shevchuk* [2]*, I.N. Voloshyna* [1] *and N.N. Gregirchak* [1]**

[1] – National University of Food Technologies, 01033, Kiev, Vladimirskaya St., 68

[2] – Institute of Microbiology and Virology, National Academy of Sciences of Ukraine, Zabolotny St., 154, 03143, Kiev, Email: tapirog@usuft.kiev.ua, fax (38-044) 220-95-55

In this study, oil-oxidizing bacteria identified as *Acinetobacter calcoaceticus* K-4, *Nocardia vaceinii* K-8, *Rhodococcus erythropolis* EK-1 and *Mycobacterium sp.* K-2, were isolated from oil-polluted soil and water samples.

The study demonstrated that keramsit being used as a bacteria-immobilizing carrier helps spead up the bacteria growth rates and the petroleum products assimilation rates. It has further demonstrated that keramsit-immobilized *Rh. erythropolis* EK-1 and *N. vaceinii* K-8 cells effectively remove trace petroleum product (up to 100 mg per liter) from water and, therefore, may be used for water purification purposes. The water purification rates were shown being variable with water flow rates and water aeration rates, and with contents of biogenic additives (sources of nitrogen and phosphorus for the bacteria).

Actual efficiency of the water cleansing process using the immobilized *Rh. erythropolis* EK-1 cells at high water-flow rates (of up to 0.68 liters per hour) and low aeration rates (under 0.1 l/l per minute), with regular injections of 0.01% diammonium phosphate, was estimated at 99.5% to 99,8 %.

Key words: oil-oxidizing bacteria, immobilization, oil-polluted water cleansing, cultivation.

Pollution of water reservoirs and soils by crude oil and petroleum products is considered one of the most negative impacts of industrial activity on the natural environment. Applying the microorganisms assimilating crude oil and petroleum products as sources of hydrocarbon and energy has proven to be one of the most promising ways of cleansing wastewater of trace petroleum products. Different matters of the oil-petroleum microorganisms' application as biological cleansers of industrial wastewaters has been addressed by many researchers and covered by the ample relevant bibliography [1-5]. Our analysis of the available research data shows that germ associations (biocenoses) and/or special microorganism cultures adapted to certain chemical pollutants are preferred when designing different biological wastewater cleansing technologies. It is further widely recognized that water cleansing process efficiency when removing crude oil and petroleum products is normally increased when the microorganisms are immobilized.

It should be noted, however, that despite the considerable research efforts to improve the existing microbiological cleansing technologies in application to oil-polluted wastewaters, the efforts still fail to solve a few top-priority problems, including, among other things, a problem of efficient water cleansing of trace crude oil and petroleum products (showing contents under 100 mg/l) at high wastewater flow rates through the media with immobilized microorganisms. The present study offers one of the possible ways to help solve the problem. Its objectives were as follows: (1) Isolate oil-oxidizing microorganisms from oil-polluted soil and water samples, and identify them; (2) Study the assimilation ability of the free and immobilized microorganisms' cells for a variety of hydrocarbons; (3) Use an experimental laboratory unit to study the efficiency of oil-polluted water cleansing process using the immobilized microorganisms' cells, for high water flow rates; and offer potential approaches to speed up the cleansing technology.

MATERIALS AND METHODS

Collecting and Isolating the Growing Oil-Oxidizing Microorganisms

The oil-oxidizing microorganisms used in the study were isolated from soil and water samples that were sampled at the Nadvirne-based Oil-processing Plant site in Ukraine. The growing cultures absorbing different hydrocarbons were obtained by inputting 0.5 g of soil or 1 ml of water sampled to the mineral medium of the following composition (g/l): Na_2HPO_4 x $12H_2O$ = 2.0; KH_2PO_4 = 1.0; NH_4NO_3 = 1.0; NaCl = 1.0; Na_2CO_3 = 0.2; $MgSO_4$ x 7 H_2O = 0.2; $MnSO_4$ x 5 H_2O = 0.02; $CaCl_2$ x H_2O = 0.01; $FeSO_4$ x 7 H_2O = 0.01. Sources of hydrocarbon used in the study included: light crude oil (with specific gravity of 0.85 g/cub. cm) in the amounts of 100 to 1000 mg/l; liquid paraffines (n-alcanes) in the amount of 0.5 vol.%; and hexadecane in the amount of 0.5 vol.%. The microorganisms were cultivated at 30°C and pH 7.0 for 5 days. A previous fermentation culture in amount of 10 vol.% was used as the inoculate for sequential passages of strains. The target growing cultures were received through 3 to 5 sequential passages of strains on the above-indicated medium.

Monocultures Isolation Procedure

The monocultures were isolated by passage of the growing cultures under the Koch assay on nutrient agar mediums including the following: meat-peptone agar medium (MPA); glucose-potato agar medium (GKA) and mineral medium containing 0.1% of crude oil and liquid paraffines. Regular testing assays [6] were used to determine the cultural, physiological and biochemical properties of the cultures, and the microorganisms were identified using [7] assay.

Monocultures Growing Procedure

The isolated monocultures were grown on the mineral medium of the above-indicated composition and Kodam [8] medium containing such hydrocarbon nutrients and light crude oil (with specific gravity of 0.85 g/cub. cm) in the amounts of 500 to 1000 mg/l; liquid paraffines (n- alcanes) in the amount of 0.5 vol.%; and ethanol in the amount of 1 vol.%. The microorganisms were grown up in flasks, each with 100 ml of medium on a shaker (at 220 rpm rotation rate) at 30°C for 16 to 96 hours. The medium was inoculated either with a 1-day-old culture grown on MPA or with an exponential-phase (16-to 18-h-old) culture grown on the mineral Kodama medium containing 0.5 vol % ethanol.

Some bacterial strains were obtained through geterophase cultivation on an oil- and ethanol- containing mineral medium with keramsite. The keramsite was preliminary grinded down to 2-3 mm size grains, put to 750-ml flasks and sterilized at 121°C for 1 hour.

Biomass quantities were determined based on the optical density of the cell-containing suspension, with the dry mass being computed using a calibration diagram and through direct weighing method. Specific germ growth rates were determined as described in article [9].

Hydrocarbon contents in the cultural liquid were determined through direct weighing method. Hydrocarbons for weighing were produced using a triple extraction by hexane, followed by the hexane extract evaporation using a rotation evaporator, at 40°C till the resultant mass showing a constant weight.

Germ Cells Immobilization Factor Identification Procedure

The study microorganisms were grown up till the end of the exponential phase on the Kodam mineral medium containing 1 vol.% of ethanol and 0.5 vol.% of liquid paraffines. 2 to 3 mm fraction of keramsite was used for immobilization of the microorganisms. A 3.5 x 22 cm test-glass was filled with this keramsite fraction up to 18 cm from the bottom, washed up by a tap water, and then a cell suspension with optical density of (0.15 – 0.2) x 10 was added to the tube. In 24 hours the cell suspension was poured off, and optical density of the solution was determined. Cells immobilization factor was determined as a ratio of the input and output suspension optical density difference to the input suspension optical density, in percentage terms.

Experimental Unit Efficiency Rating Procedure for the Oil-Polluted Water Cleansing Process

A drawing of the experimental laboratory unit designed for oil-polluted water cleansing technology development purposes is given in Figure 1. Efficiency of the water-cleansing column was rated taking into assumption technical specifications of a real water-cleansing unit installed at one of the oil-processing industrial plants, including:

- Water flow rate of 4 cubic meters per hour;

- Keramsite carrier column of 1 meter in height;
- 10 meter effective diameter of the unit;

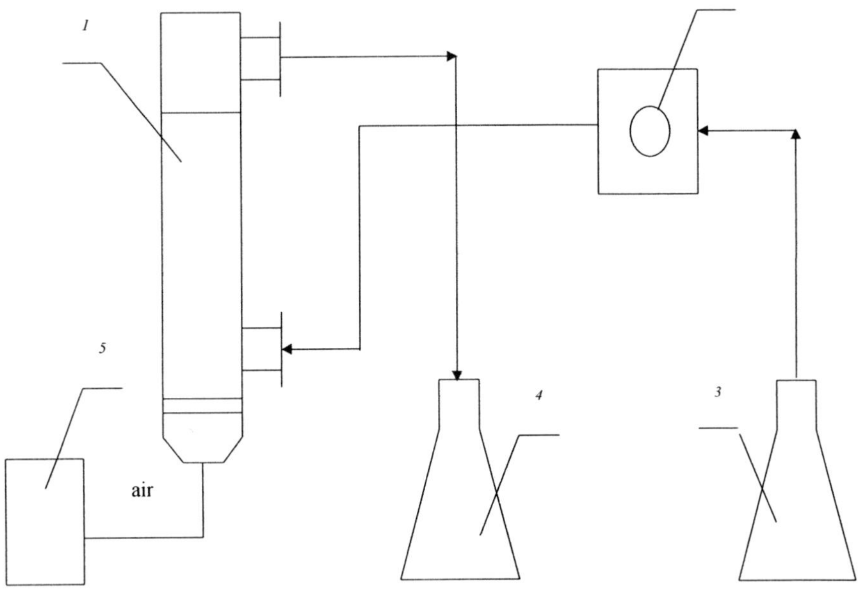

Figure 1. Flow chart of the experimental laboratory unit for oil-polluted water cleansing using immobilized microorganisms: (1) keramsite-filled column; (2) peristaltic water pump; (3) oil-polluted water containing flask; (4) flask with cleansed water; (5) micro-compressor set.

This industrial unit was chosen for comparative estimates of the laboratory unit efficiency rates, to ensure the best process efficiency at high water flow rates, to improve the currently low water cleansing efficiency of the industrial process.

Volume of the keramsite column is estimated as $V = S \times H = \pi d^2/ 4H = 3.14 \times 10^2/ 4 \times 1 = 78.5$ cub.m. Efficiency rate of the column is estimated at $Q = S \times V = \pi d^2/ 4V = 3.14 \times 10^2/ 4 \times 4 = 314$ cub.m/ h. Flow rate is estimated as $D = Q/V = 314/ 78.5 = 4\ h^{-1}$.

Total volume of the keramsite column in the laboratory unit makes up $V = 3.14 \times 0.035^2/ 4 \times 0.18 = 0.00017$ cub.m. Therefore, efficiency rate of the laboratory unit is determined as $Q = V \times D = 0.00017 \times 4 \times 1000 = 0.68$ l/h.

Cell Immobilization Procedure for the Laboratory Unit

The keramsite column was sterilized at 121°C for 1 hour. All further process to use the keramsite-immobilized oil-oxidizing bacteria for water cleansing at the experimental laboratory unit was conducted in non-sterile conditions at normal indoor temperatures of 18 to 20°C.

Prior to cells immobilization in the laboratory column, the keramsite filler was saturated with crude light oil having a density of 0.85 g/cub.cm. Water containing 100 to 500 mg/l of oil was put through the aerated column and sampled every 4 hours, the 200 ml samples being taken on the inlet and outlet of the column to test oil contents through a direct weighing method. The process keramsite column was saturated with the above oil until oil contents in the input and output water reach the same level.

A culture grown by the end of the exponential phase on the Kodam medium including 1 vol% of ethanol was used to immobilize *Rhodococcus erythropolis* EK-1 cells on the oil-saturated keramsite; and the Kodam medium with 0.5 vol%. of paraffines was used to immobilize *Nocardia vaceinii* K-8 cells. Cell suspension having the optical density of 0.15 x 10 was injected in the process column by a NP-1M type peristaltic pump at the rate of 0.1 h^{-1}, with re-circulation of the process water. Cells were immobilized for 24 hours in the aerated column, at the air flow rate of 0.05 l/l per minute.

Oil-Polluted Water Cleansing at the Experimental Laboratory Unit

After the cells were immobilized in the aerated column, tap water containing 100 to 250 mg/l of crude light oil having a density of 0.85 g/cub.cm, was pumped through the column. 0.01 vol.% to 0.1 vol.% of diammonium phosphate $(NH_4)_2$ HPO_4 were periodically added to the water in the process. Water cleansing efficiency was estimated for the water flow rates of 0.3, 0.5 and 0.68 liters per hour. Air flow rates were maintained in the process at the level of 0.05 to 0.10 l/l per minute using an AEI-3 type micro-compressor set. The process water was sampled every 6 to 8 hours, the samples of 200 ml each being taken on the inlet and outlet of the process column, followed by oil contents determination using the direct weighing method, the resulting data being supported by infrared (IR) spectroscopy analyses using an AN-1 type petroleum product analyzer of Russian production. Continuous process cycle of the column was set at 30 days.

RESULTS AND DISCUSSION

In the study process, 10 growing cultures were collected from 12 samples of oil-polluted soils and water, and the cultures assimilated all hydrocarbon additives within 3 days of cultivation process. The follow-up studies demonstrated that the growing cultures are composed of 3 to 4 types of bacteria. Research efforts to assess growth rates of the isolated monocultures on different oil-, paraffin- and hexadecane-containing media have demonstrated that most of the monocultures were less active on the media as compared to the associations the cells were isolated from. As a result, four bacterial strains were isolated and identified, using a set of morphological, cultural, physiological and biochemical indicators, as *Acinetobacter calcoaceticus* K-4, *Nocardia vaceinii* K-8, *Rhodococcus erythropolis* EK-1 and *Mycobacterium sp.* K-2. Study data showing these bacterial strains cultivation process on different media are given in Table 1.

Table 1. Monoculture growth indicators for Kodam medium and for different hydrocarbon nutrients

Monocultures	Substrates								
	Media Oil, 1000 mg/l			Paraffines, 0.5%			Ethanol, 1%		
	Biomass, g/l	μ_{max}, h^{-1}	Media assimilation rate, %	Biomass, g/l	μ_{max}, h^{-1}	Media assimilation rate, %	Biomass, g/l	μ_{max}, h^{-1}	Media assimilation rate, %
Rhodococcus erythropolis EK-1	0.40	ND	60	1.45	0.06	100	3.20	0.10	100
Acinetobacter calcoaceticus K-4	0.25	ND	45	0.55	0.03	75	3.50	0.12	100
Nocardia vaceinii K-8	0.35	ND	60	1.30	0.05	100	0.65	ND	40
Mycobacterium sp. K-2	0.35	ND	60	0.95	0.04	85	0.50	ND	35

ND stands for "not determined."

Table 2. Monoculture growth indicators for Kodam medium in the presence of keramsite

Monocultures	Source of hydrocarbon	Process indicators						Cells immobilization rate, %
		Without keramsite			With keramsite			
		Biomass, g/l	μ_{max}, h^{-1}	Media assimilation rate, %	Biomass, g/l	μ_{max}, h^{-1}	Media assimilation rate, %	
Rhodococcus erythropolis EK-1	Ethanol	3.20	0.10	100	3.65	0.14	100	85
	Oil	0.40	ND	60	0.45	ND	80	
Acinetobacter calcoaceticus K-4	Ethanol	3.50	0.12	100	3.60	0.16	100	70
	Oil	0.25	ND	45	0.25	ND	65	
Nocardia vaceinii K-8	Oil	0.35	ND	60	0.35	ND	80	80

Note: Cultures produced at the exponential growth phase grown on the Kodam medium with 0.5% of ethanol, were used as inoculate.
ND stands for "not determined."

Every of the studied strains is able to grow on a hydrocarbon medium, though the liquid paraffin absorption rates were notably higher than that for crude oil (75% to 100% vs. 40% to 70%, respectively). It should be noted that *Acinetobacter calcoaceticus* K-4 and *Acinetobacter calcoaceticus* K-4 and *Rhodococcus erythropolis* EK-1 strains showed most high growth rates and biomass accumulation rates when cultivated on an ethanol-containing medium. It may be also pertinent to note that only the ethanol-containing media triggered formation of a homogenous cell suspension that is considered an important condition for cells immobilization. When paraffines were used as a source of hydrocarbon, cells were often found to form conglomerates, with partial or full flotation in the process.

Cultivation of *Acinetobacter calcoaceticus* K-4 cells on the oil-polluted media resulted in coalescence of oil with formation of 3 to 4 mm oil balls when oil contents were at the level of 500 mg/l, and formation of oil discs having 7 to 8 mm in diameter and thickness of 1 to 2 mm. The process of oil coalescence may be due to a synthesis of off-cellular metabolites have some surface activity. This question will be addressed in our further studies. It may be worth noting that the coalescence-triggering effect of these bacterial strains and/or their metabolites on petroleum products may be of significant interest for a wide variety of industrial applications, including cleansing of oil tankers, oil-transmission pipelines, tank barges and different vessels of crude oil and petroleum products.

Our experiments have demonstrated that when three other strains were grown on an oil-containing medium, up to 30 vol.% of oil were absorbed on glass of the test flasks, and hydrocarbons were extracted from flask walls after the cultural liquid were poured off. This may be indicative of the fact that trace oil found in the cultural liquid (30% to 40%, see Table 1) may be due to its washing from the flask walls.

Table 1 gives cultivation process data for the studied strains when MPA-containing inoculate was used. It may be pertinent to emphasize the fact that the oil-oxidizing ability of the strains grown up on ethanol-containing mineral media showed virtually no variability.

Our further experiments demonstrated that the studied *A. calcoaceticus* K-4, *N. vaceinii* K-8 and *Rh. erythropolis* EK-1 strains were able to grow and propagate in the presence of keramsite (Table 2). Moreover, the studies proved both the maximum specific germ growth rates and (in some cases) the biomass accumulation rates being on the rise in the process. It is of particular interest that the amounts of oil absorbed in the presence of keramsite increased by 20% to 25%, while the biomass accumulation rate remained virtually unchanged. This fact appears to be due to the process of oil absorption by keramsite. Immobilization rates for the studied bacterial strains on keramsite made up 75% to 85% (Table 2).

For further studies of the oil-polluted water cleansing process using keramsite-immobilized bacterial cells, we selected *N. vaceinii* K-8 and *Rh. erythropolis* EK-1 strains as the ones that demonstrated the highest growth rates on hydrocarbon media as well as the highest cells immobilization rates on keramsite.

To offset the effect of crude oil absorption on the carrier – as we registered notable absorption (up to 20%) when cultivated the *N. vaceinii* K-8 and *Rh. erythropolis* EK-1 strains on the oil-containing media in the presence of keramsite – we saturated keramsite with oil prior to cells immobilization in the experimental unit. It is also well known that *N. vaceinii* and *Rh. erythropolis* cell walls contain mycolic acids that are believed to contribute to hydrophobic properties of the cells [7]. In our opinion, these properties may provide for more effective immobilization of these cells on the oil-saturated keramsite.

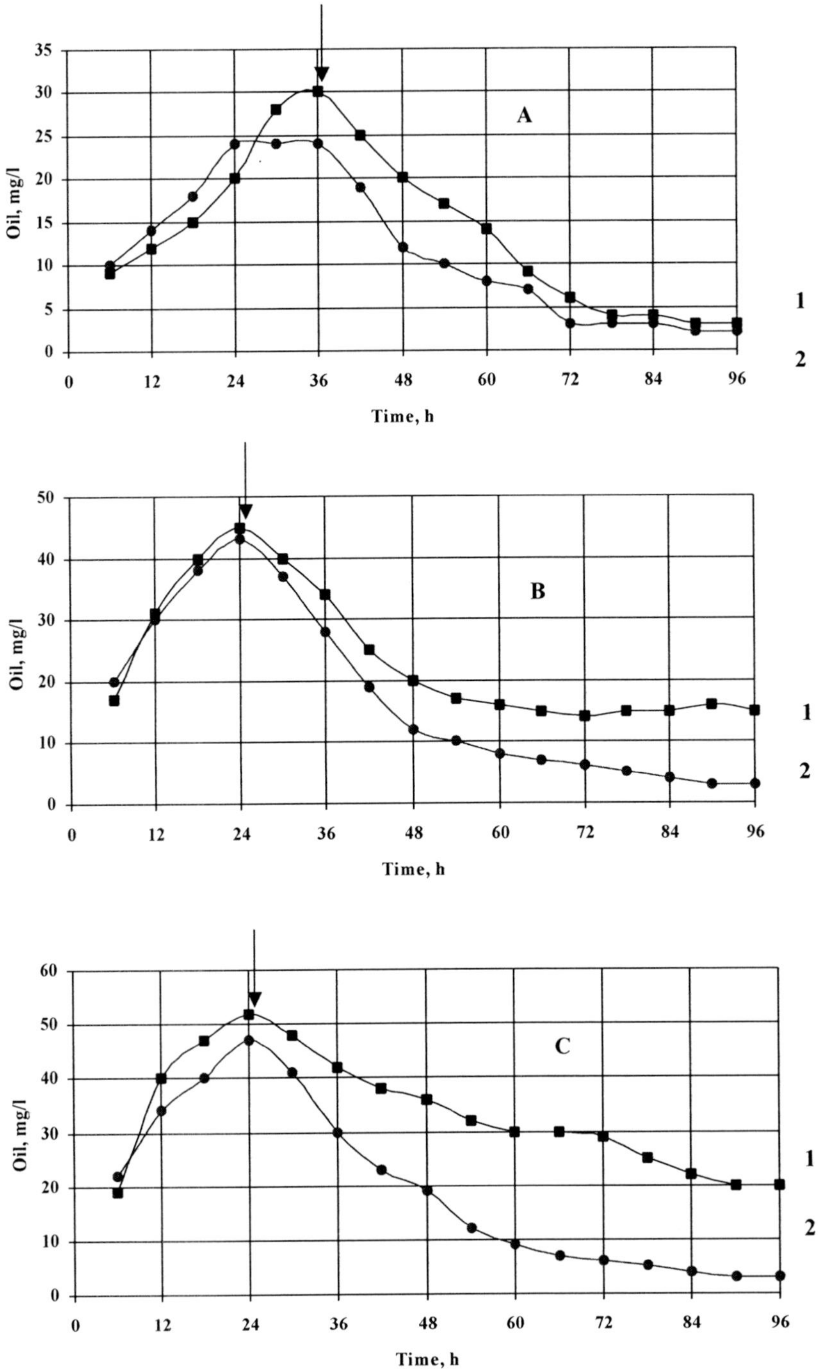

Figure 2: Influence of mineral nutrients and oil-polluted water flow rates on the water cleansing efficiency for the immobilized *N. vaceinii* K-8 (1) and *Rh. erythropolis* EK-1 (2) cells:
Flow rates (liters per minute): A = 0.3; B = 0.5; C = 0.68.
Initial oil content in water = 100 mg/l.
↓ Added diammonium phosphate (0.1%).
Oil content was determined using a direct weighing method.

After the cells were immobilized, water containing 100 mg/l of crude oil was pumped through the process column at different flow rates (Figure 2). As demonstrated by the data presented in Figure 2, we always registered in 24 to 36 hour of the process, largely irrespective of the flow rates, a gradual rise of oil contents in the output water to as much as 30 to 50 mg/l. It is worth noting that trace concentrations of crude oil in the output water tended to rise with an increase of the flow rate, from 30 mg/l for the flow rate of 0.3 l/h to 50 mg/l for the flow rate of 0.68 l/h.

It is widely believed that an application of bacterial cultures to absorb oil pollutants may be efficient only when mineral nutrients are added in the process. It was first reported as far ago as in middle 50ies by Izurova [10] that nitrogen-, phosphorus- and potassium-containing additives notably speed up oil-degradation processes. Presence of nitrogen and phosphorus is considered among necessary conditions for oil-degradation process when oil tankers are cleansed using *Acinetobacter calcoaceticus* RAG-1 strain [11]. Some studies have demonstrated that a single compound being used as an additive (such as Na_2HPO_4, NH_4NO_3 or KNO_3), even in highest concentrations, fails to significantly increase the oil degradation process [12]. However, a combination of phosphate additives always resulted in notable rise of the oil absorption rates.

Our study showed that the output oil contents in the water leaving the process column (Figure 2) fell when 0.1% of diammonium phosphate was added to water. In case of the *Rh. erythropolis* EK-1 strain, for instance, oil contents went down to 3 to 4 mg/l in the period of 72 to 96 process hours. As for the case of *N. vaceinii* K-8 strain, the same oil contents in the output water were registered only at the lowest flow rate of 0.3 l/h. When the flow rate was higher than that, its trace oil contents were varying at the level of 15 to 20 mg/l. Figure 2 presents trace oil data provided by the direct weighing method. Data of the IR spectroscopy, however, show trace oil contents in the output water one or two degrees of magnitude lower as compared with the direct weighing method: specifically, in case of *Rh. erythropolis* EK-1 strain, irrespective of the water flow rate through the process column, they stay at as little as 0.05 to 0.10 mg/l.

It may be important to mention that the drop in trace oil contents on the outlet of the process column when 0.1% of diammonium phosphate was added was associated with notable rise of water turbidity. Microbiological assay showed presence of the studies bacterial cells in the output water, that may be indicative of propagation of the bacterial cells in the presence of 0.1% of diammonium phosphate and their release from the carrier. To avoid that effect, concentrations of the biogenous additive in the further experiments were kept under 0.01%. In these conditions, the experimental laboratory unit showed efficient and stable performance for the period of 7 days: specifically, at the flow rate of 0.68 l/h, trace oil contents in the output water varied within 2 to 3 mg/l for the studied *Rh. erythropolis* EK-1 strain and around 12 to 16 mg/l for the *N. vaceinii* K-8 strain, according to the data of weighing method. Oil contents determined by the IR spectroscopy made up 0.05 to 0.09 mg/l and 0.2 to 0.4 mg/l for the studied *Rh. erythropolis* EK-1 strain and *N. vaceinii* K-8 strain, respectively. Neither of the bacterial strains was found in the output water. However, the output water turbidity marked a notable rise on day 8 of the experiment, apparently due to the cells propagation in the process. Our further experiments demonstrated that 0.1% of diammonium phosphate should be added to oil-polluted water on a regular basis: during 6 to 7 days, followed by a break of 1 or 2 days.

Increased aeration up to the level of 0.1 l/l of water per minute resulted in the rise of water cleansing efficiency for the both studied strains (Figure 3). However, the trace oil contents in the output water for *N. vaceinii* K-8 strain were still 2.5 to 3.5 higher that that for *Rh. erythropolis* EK-1 strain.

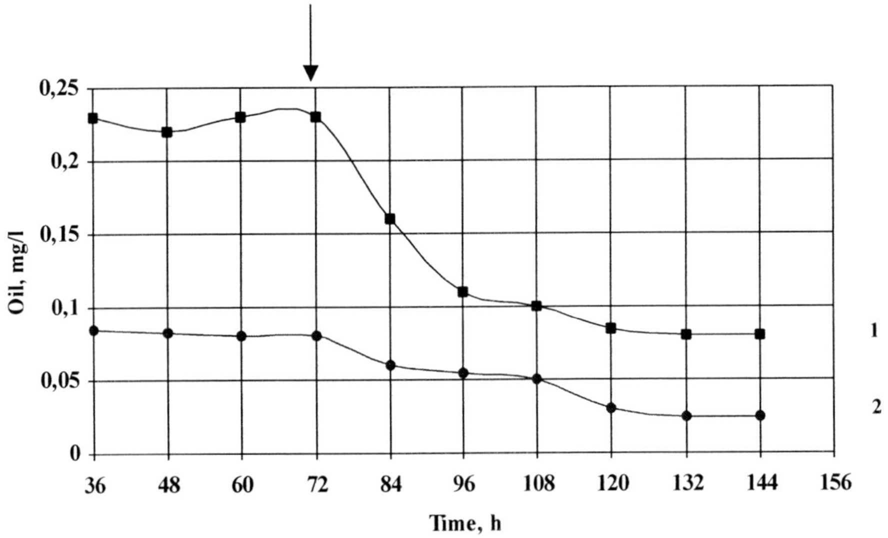

Figure 3. Aeration rate influence on the water cleansing process for the immobilized *N. vaceinii* K-8 (1) and *Rh. erythropolis* EK-1 (2) cells:
Initial oil content in water = 100 mg/l; water flow rate = 0.68 l/h; initial aeration rate = 0.05 liters of air per liter of water per minute;
↓ Increasing aeration rate up to 0.1 liter of air per liter of water per minute.
Oil content was determined using IR spectroscopy.

When initial oil content in water was raised from 100 mg/l to 250 mg/l (with water flow rate and aeration rate being maintained at the levels of 0.68 l/h and 0.1 l/l per minute, respectively), we registered sagging of the water cleansing efficiency for *N. vaceinii* K-8 strain that came down to 90% at most, while for the *Rh. erythropolis* EK-1 strain it stayed virtually at the same level of 99.0% to 99.5%.

This high efficiency level of water cleansing process for the immobilized *Rh. erythropolis* EK-1 strain was registered for 30 days of uninterrupted process at the experimental laboratory unit, in non-sterile conditions.

According to a Regulation of July 6, 1993 now in effect in Ukraine, petroleum product contents in industrial wastewaters incoming in municipal sewerage systems and water reservoirs shall not exceed 4.4 mg/l in case of a municipal sewerage and 0.3 mg/l when they are discharged in water reservoirs.

As a result of the study, we collected from oil-polluted soil and water samples and identified important bacterial cultures capable of oil assimilation in contents fairly close to the permissible oil contents in wastewaters, and facilitating crude oil coalescence in "oil-in-water" systems. Furthermore, the study demonstrated that the studied bacterial culture cells are able to grow and propagate in the presence of keramsite, and immobilization rates of the cells on this carrier come to as much as 75% to 85%. Under experimental conditions of the study, we demonstrated good potential of the water cleansing process for trace oil contents of 100 to 250 mg/l when keramsite-immobilized bacterial cells comprise a cleansing agent, and the process is run at high water flow rates (up to 0.68 l/min) and low aeration rates (under 0.1 liters of air per liter of water per minute). It was further shown that the water cleansing efficiency in case of oil pollution was notably boosted by biogenic additives: when 0.1% of diammonium phosphate was added on a regular basis, the water cleansing efficiency made up as much as 99.5% to 99.8%.

REFERENCES

1. M.B. Bitteyeva, B.G. Murzakov, G.M. Morshakova et. al.: RF Patent, 2014286, 1994. (in Russian).
2. G.A Kozhanova: RF Patent, 2031860, 1995. (in Russian).
3. B.G. Murzakov, G.M. Morshakova, L.N. kapotkina: RF Patent, 2053204, 1996. (in Russian).
4. L.V. Ogurtsova, T.N. Morozova, Y.V. Zhdanova: RF Patent, 2129604, 1999. (in Russian).
5. G.G. Yagafarova, A.K. Safarov, Y.G. Ilyina et al.: *Prikladnaya Biochemiya i Microbiologiya., **38** (5),* 518 (2002). (in Russian).
6. F. Gerhardt (Ed.): Methods of General Bacteriology. *Mir, Moscow, 1984,* **3**, 264 p.
7. Bergy's Identificator of Bacteria. 9th Edition. *Mir, Moscow, 1997,* **1**,**2**. 800 p.
8. T. Kodama, T. Nakahara, T. Omori et al.: Microorganisms growth on C_1-compounds. Summary report. NTBI AN SSSR, *Pushino*, 1977, p 213. (in Russian).
9. S.J. Perth: Fundamentals of Cultivating Microorganisms and Cells. *Mir, Moscow, 1978,* 333 p. (in Russian).
10. A.I. Izcyurova: *Hygiene and sanitary, **6** (5),* 15 (1956). (in Russian).
11. Gutnick D.L., Wise G.S.: *Biopolimers, **26**,* 223 (1978).
12. Atlas R., Bartha R.: Biotechnol. and *Bioeng, **14*** (3), 309 (1972).

Chapter 14

The Use of Immobilized Microorganisms and Aminomethylnitroamines for Intensification of Bioremediation of Oilcontaminated Soil

T.V. Sakhno [1,2]***, A.A. Kolesnik*** [3]***, V.M. Kurashov*** [1,2] ***and V.S. Gavrilov*** [3]

[1]Kazan State Technological University, Kazan, 420015, Russia

[2]Research and Production Centre "Ecological Consortium", Kazan, Russia,

[3]Central Directorate of Natural Resources and Preservation of Environment of the Ministry of Natural Resources of Russia in the Tatarstan Republic, Kazan, 420059, Russia,

Abstract

Biopreparations were developed including immobilized microbiological and intensifying addings. These compositions are developed for cleaning the territories and waters of oil contaminations when the oil concentration exceeds the maximum permissible concentration at which the vital functions and effective work of bacteria are possible. We used the sorbents of inorganic and organic origin to immobilize the microorganisms, for example, phytogenic farm waste (buckwheat husk, oats husk, wellmilled chaff of grain crops), or waste from timber, resin and paper industry (sawdust, lignin). The Proxanol oxygen transferor (P-268) and aminomethylnitroamine, 3,3`-ethylene-bis-(1,5-dinitro-1,3,5-triazacycloheptane) (AMNA) were used as intensifying addings. The joint use of immobilized microorganisms, Proxanol and AMNA allows decontamination of soils of oil and oil products having a concentration 4.5 times above the limit, and it also allows decreasing of the decontamination process duration.

Key words: bioremediation, immobilization, oil, microorganisms, Proxanol, oxygen transferor, aminomethylnitroamine.

Introduction

It is known that the intensity of self-purification of soils from oil is determined by microrganisms. Hydrocarbon-degrading microorganisms are a permanent component of soil and water microbiocenoses. But owing to emergencies the concentration of oil and oil products are in excess of the limiting values for the vital activity of microbiocenosys. In this connection, among the measures focused on the increasing the productivity of biopurification of lands and reservoirs from oil and oil products, immobilized microorganisms seem to hold the lead. In recent years the biological adsorption method depending on the employment of various carries with well-developed sorption surfaces for microorganisms has become very popular [1]. Immobilization of cells on the carriers observably increases the effectiveness of biosynthesis, biocatalysis, and biodegradation [2,3,7]. The effectiveness of oil biodegradation by microorganisms at high level of contamination is restricted by the availability of oxygen. Various stimulating substances are used in order to diminish an influence of the limiting factors on the growth and productivity of microorganisms [4,5]. The Proxanoles surface-active substances consisting of ethylene and propylene oxides block copolymers are very promising in this regard [6].

The purpose of the present work is the mutual using of immobilized microorganisms with oxygen transferors and AMNA in external conditions when high concentrations of oil spills lead to the lowering of oxidizing capability of hydrocarbon-degrading bacteria.

Works in this area promote the restoration of natural potentials and recycling of agricultural objects, forests and bodies of water back into the national economic circulation.

The obtained results are used in oil decontamination of the territories in the Republic of Tatarstan.

METHODS AND MATERIALS

The subject of our investigations is the microbial preparation Putidoil manufactured on the base of oil-degrading bacterial culture Pseudomonas Putida-36 and "Destroil" based on bacteria Acinetobacter genus. Oxygen transferors Proxanole-268 and aminomethylnitroamin 3,3-ethylen – bis (1,5 dinitro-1,3,5 –triazocycloheptane) (AMNA) together with intensifying additions and oxygen transferors were used as intensifying additions. Carriers of various types were used as sorbents for immobilization of destroyer microorganisms, including ones of inorganic, organic, and biogenic origin [radiochemically destructured hen's feather awn (HFA)].

Immobilization was carried out by mingling sorbents with liquid culture for 1 h and drying the samples at 30-35^{0}C. The destructive capability of the microorganisms was checked by the weight method through the residual content of oil and oil products in the medium after processing three-times with hexane, tetrachlorated carbon, and chlorophorm.

Experiments for cleaning oil-contaminated soil were carried out on the territory of Vostochno-Suleevskii Romashkinskii's deposit of Republic Tatarstan on two plots. Both plots have a waterproof clay.

The cause of the pollution on the first plot (№1) was a break in an oil pipeline; the square of pollution was 100%. The oil concentration was 110 g/kg; the depth of oil penetration was 30 cm. The second plot (№2) was spoiled by oil products due to the breaks of pipelines and accident escapes of gas-liquid mixture from the wells. Oil concentration was 80 g/kg in soil, sodium ions content 0.03%, chloride-ions 0.06% (according to the salts' analysis of water extract from soil).

The first plot was divided into 21 micro plots, the second one – into 40 micro plots having each 2x2 m size each, located 10 m from each other. Seven variants in 3 replications were prepared. On the first plot (N1) three micro plots were reference ones without intensifying additions and microorganisms. In model variants, 1.5 g of the biopreparation Putidoil was dispersed per square metre as an aqueous solution without oxygen transferor and with it (P268), or as immobilized cells on the sorbents with and without oxygen transferors. The carriers of inorganic (swelled perlite (SP)) and organic (radio chemically destructed hen's feather awn (HFA)) origin were used as the sorbents for microorganisms' immobilization on the plot N1.

The plot N2 was ploughed up to 40 cm depth and divided into 6 test and 34 plots of $4m^2$ each. The surface of the first two test micro plot was not treated at all. The surface of four other test plots (N3, N4, N5, N6) was treated by bacteris Pseudomonas Putida, Acinetobacter genera, oxygen transferor P-268, and 3,3-ethylen – bis (1,5 dinitro-1,3,5 – triazocycloheptane), respectively.

Proxanoles that are uninogenic surface-active substances consisting of ethylene and propylene oxides block copolymers, were used as oxygen transferors.

In pilot variants 1.5g of boipraparations "Putidoil"or "Destroil"were injected per 1 m^2 of soil surface. Those biopreparations were immobilized either by vegetable wastes of agriculture industry, or by wastes of timber and paper industry: both in the presence of oxygen transferors (P-268) or AMNA, and without them. Per $1m^3$ of oil-contaminated surface 10 g of oxygen transferor or AMNA an 1kg of modified composition were injected. 25 types of modified immobilized microbioligical compositions were used:

1/ the wastes of agriculture and vegetable origin: buckwheat husk (BH), oats husk (OH), crumbled straw of cereals (CSC);
2/ wastes of timber or paper industry : acid lignine (AL) sawdust (S).

Intensifying additions "Proxanol-268"and methylnitroamin were injected in the amount of 0.1g per $1m^2$ of soil.

Model and reference micro plots were fenced with viniplast frames 65 cm high emplaced to a depth of 40 cm. Plots were ploughed beforehand. After that a mineral feeding was performed at a rate of 0.8 g of ammonium phosphate plus KNO_3 per square metre of soil. The soil was made friable to a depth of oil penetration once a week. Humidity was created by adding fresh water up to the soil's total hydrocapacity. Ten samples (10 g each) were withdrawn from the upper layer of soil after 10-20 days during 200 days on the plot N1 and during 90 days on the plot N2.

They were thoroughly averaged out and were subjected to a study of the biodegradation of oil spills through the residual content of oil in the soil by means of the weight method. The effectiveness of biodegradation (E) was obtained by using the formula:

$$E = [(C_0 - C_x) / C_0] \cdot 100\% \qquad (1)$$

where C_0 is the initial concentration of oil in the medium, g/kg, and C_x is the concentration of oil in the medium during biodegradation, g/kg.

RESULTS OF PILOT-INDUSTRIAL TESTS И DISCUSSIONS

Biodegradation process in soil is poorly known. Difficulties appear, in part, due to the variable composition of soil. Implementation of pure preparation Pseudomonas putida-36 preparation for soil remediation is effective when oil concentration is 20-25 g/kg. Higher concentrations lead to inactivation of oxidizing activity of biopreparation and then to its death because of inhibiting character of hydrogencarbon substrate. Using of immobilized microorganisms and oxygen transferors provides an opportunity for applying hydrocarbon-degrading bacteria for oil biodegradation at a concentration of oil exceeding the limiting one [6,7]. A character of changing of oil concentration and the effectiveness of oil biodegration during microbial cleaning from oil contaminations on the plot N1 using immobilized compositions both with and without a presence of oxygen transferors is demonstrated in the Table.

Table. Oil biodegradation in soil on the plot N1 by immobilized microorganisms Pseudomonas Putida using oxygen transferor.

Ps. Putida immobilized on the sorbents

Time, Days	SP		SP + P268		HFA		HFA + P268	
	C, g/kg	E, %	C, g/kg	E, %	C, g/kg	E, %	C, g/kg	E, %
0	110	0	110	0	110	0	110	0
20	110	0	109	0,1	106	3,6	104	5,4
40	107	2,7	105	4,5	103	6,3	100	9,1
60	102	7,2	100	9,1	95	13,6	87	20,9
80	91	17,3	83	24,5	72,5	34	62,5	43,2
100	72,5	34	60	45,45	47	57,3	32,5	70,5
120	54	50,9	37,5	65,9	23	79	12	89
140	33	70	16	85,45	4,5	95,9	2	98,2
160	17	84,5	2	98	2	98	0,75	99
180	5	95,5	2	98	0,75	99	0,75	99
200	2	98	0,75	99	0,75	99	0,75	99
220	0,75	99	0,75	99	0,75	99	0,75	99

It can be seen from the Table that the increasing of oil spills concentration (C) to 110 g/kg in chernozem seems to inhibit hydrocarbon-degrading microorganisms Pseudomonas putida of Putidoil biopreparation both without oxygen transferors, and with the P268 oxygen transferor. A spreading the active biopreparation Putidoil upon HFA and SP carriers was found to breaks down the oil up to the residual concentration of 2 g/kg, and 17 g/kg, respectively; in the presence of oxygen transferor P268 – up to 0.75 g/kg and 2 g/kg, which corresponds to the degree of purification of 9% and 98%.

Experimental data show (see Table) that lowering oil concentration up to the limiting value for technogenic objects (0.75 g/kg, the degree of biodegradation is 99%) using HFA and SP without oxygen transferor achieves after 180 and 220 hours, respectively, while mutual action of oxygen transferor P268 and immobilized carriers reduce time to 160 and 200 hours, respectively.

On the plot N2 on the test micro plots the oil concentration has not been changed during 90 days (80 g/kg), as well as the results of the analyses of water extract of soil (chloride-ions 0.06%, sodium ions 0.03%), and an anmount of bacteria in soil.

The oxidizing capability of hydrocarbon-degrading bacteria for microorganisms Pseudomonas putida 36 of "Putidoil" and Acinetobacter of "Destroil" is the same.

Hudrocarbondegrading microorganisms Pseudomonas putida 36 ("Putidoil"), bacteria of Acinetobacter genus ("Destroil") seem to degrade oil on 63% in the presence of AMNA (the residual oil content (ROC) in soil – 29g/kg) during 90 days, and in the presence of P-268 – 49% (ROC – 40.8g/kg). Immobilization of Pseudomonas putida or Acinetobacter on CSC, BH, or OH during 90 days leads to oil biodegradation on 65% (ROC – 38g/kg), in the presence of P-268 and AMNA – on 99.1% (ROC – 720 mg/kg), and 99.7% (ROC – 240mg/kg), respectively.

During 90 days biopreparations on the base of bacterial strain Pseudomonas putida –36 ("Putidoil") or Acinetobacter genus ("Destroil") seem to degrade oil in different rate depending upon the immobilized composition:

- the mentioned above bacteria immobilized on acid lignine (AL) lead to oil biodegradation on 70% (ROC 24 g/kg);
- Immobilized on sawdust – on 53% (ROC 38%)
- immobilized on AL in the presence of P-268 and AMNA – on 99.5% (ROC 400 mg/kg), and 99.8% (ROC 160 mg/kg), respectively.
- immobilized on sawdust in the presence of P-268, and AMNA, on 98% (ROC 1.6g/kg), and 99.8% (ROC 160mg/kg), respectively.

On the pilot plots the soil fertility was restored after 90 days. Oil concentration on pilot plots was lowered till limiting concentrations and amounted to 250-750 mg/kg. The chlorine content in water extracts was lowered till 0.004%-0.02%, sodium content – till 0.005%-0.03%. Bacteria quantity was raised up to 10.6-11 millions per 1 g of soil.

Thus, the use of intensifying modified additions proved to shorten the biodegradation time of oil hydrocarbons and enhance the oil degradation effectiveness up to 99.8%. Fixing microorganisms on the carriers together with the oxygen transferors and intensifying additions intensify the oxidizing activity of the destroyer microorganism Pseudomonas putida-36 and Acinetobacter and also make it possible to work in emergency situations with high n of 80-110 g/kg concentration in soil. Our results demonstrate the wisdom of mutual using of immobilized microorganisms and oxygen transferors for cleaning both soil and water areas from oil and similar pollutants.

So, firstly, the use of modified immobilized biopreparations allows biodegradation process to be shortened and increases the effectiveness of biodegradation as 99.8%. Secondly, fixation of microorganisms on carriers makes it possible to apply the bacterium Pseudomonas putida-36 and Acinetobacter together with intensifying additions and oxygen trnsferorss oil biodegradation in emergency situations with oil concentration 4.5 times higher than the limiting concentration for the use of unimmobilized culture.

REFERENCES

1. N.I. Pavlenko, V.V. Izzheurova, L.M. Khenkina et al.: Using of microbes-destroyers in cleaning sewage from oil products. *Microbiol. J.*, **53**, 91-95 p., (1991).
2. N.V. Doronina, V.A. Ezhov, YU.A. Trotsenko: Aerobic biodegradation of formaldehyde, methanol and methylamine by immobilized cells. *Prikl. Biokhimiya i mikrobiologiya J.*, **33**, 162-165 p., (1997).
3. H.M. Ehrhard and H.G. Rehm, Semicontinuous and continuous degradation of phenol by Pseudomonas Putida P 8, adsorbed on activated carbon. *Appl. Microbiol. and biotechnol.*, **30**: 312-317 p., (1989).
4. Dong Jin Kim and Ho Nam Chang, *Biotechnology letters*, **12**: 289-294 p., (1990).
5. G. Sakellaris and P. Gikas, *Biotechnology letters*, **13**: 217-222 p., (1991).
6. T.V.Sakhno, V.M. Kurashov, V.M. Emelyanov *Intensification of Removal of Hydrocarbon Contamination from Water and Soil Using Oxygen Transferors. Remediation Engineering of Contaminated Soils*, 2nd Edition. of Marcel Dekker, New York, 2000, Chapter 13, 279-289 p.
7. T.V.Sakhno, N.I. Philinova, V.M. Kurashov *Using Immobilized Microorganisms to Intensify the Removal Hydrocarbon Contaminations from Water and Soil. Bioremediation of Contaminated Soils*, 2nd Edition of Marcel Dekker, New York, 2000, Chapter 24, 373 - 382 p.

Chapter 15

MICROBIAL PREPARATION RHODOBEL FOR DECONTAMINATION OF SOIL POLLUTED WITH OIL AND DERIVED PRODUCTS

A.S. Samsonova, Z.M. Aleschenkova and N.F. Syomochkina
Institute of Microbiology, National Academy of Sciences, 220141, Minsk, Belarus

ABSTRACT

Rhodobel is the first microbial preparation where specially selected strains degrading oil and processing products serves as the active principle. The degree of oil decomposition reaches in 3,5 months 100% for aliphatic components, and 95% for aromatic components. Rhodobel application ensures simple soil remediation procedures and complete ecological safety.

Key words: microbial degraders, degradation, pollution, oil, bioremediation, introduction

Pollution of the environment with oil and oil-processing products is a complex multifaceted ecological problem. Natural remediation of the environment exposed to oil pollution proceeds in biocenoses made up by association of microorganisms, Protozoa and worms.

Nowadays elaboration of methods intensifying decomposition of oil and oil-processing products is streamlined along 2 main trends and incorporates either on-site soil remediation by activation of indigenous microflora and by introduced strains-degraders or treatment of polluted soils in bioreactors. Each case sets a problem of desirability and efficiency of the introduction into the native ecosystem of non-toxic and non-pathogenic microbial strains or biological recovery of contaminated soils in specially equipped devices. In each variant acceleration of oil breakdown in soil is achieved by degradative activity of the introduced and indigenous microbial species.

Domestic and foreign experience evidences that in emergency situations requiring detoxification of large amounts of spilled oil products at a relatively small area, application of microorganisms may be justified economically [1]. Introduction of an Ecoil preparation based on oil-oxidizing bacteria Mycobacterium flavescens EX 91 into soil polluted with oil and derived products at 1% contaminant concentration results in increased mycobacterial titer from 10^7 cells/cm^3 to 10^9 cells/cm^3 [2]. Treatment of industrial soil sites at the territory of Surgut oil-gas concern with bacterial preparation Bacispecin based on native soil strain Bacillus sp. 739 accelerated process of oil biodecomposition. In experiments with tested biopreparation 52–56% of oil was split upon 2.5 months of observation when compared to the control – untreated oil-polluted soil (pollution level – 10% oil to overall soil weight) [3]. To remediate black earth plots from oil spill at one of Tatarstan oil fields, method of activating indigenous soil microflora was applied and, as an alternative, Devoroil preparation was introduced into soil. Contamination rate varied from moderate 60 m^3/ha to heavy (180 m^3/ha). After one year of investigations maximal oil disruption (60–94%) was recorded in the variant envisaging introduction of hydrocarbon-oxidizing microorganisms [4]. Application of bacteria Rhodococcus erythropolis E 15 onto tundra soil sites contaminated with oil levels 1 kg/m^2, ensured reduction of hydrocarbon load 5-fold upon 5 weeks of monitoring [5]. Efficient oil decomposition in soil was demonstrated using microbial preparation made up by strains of rhodococci and pseudomonades immobilized on vermiculite. Its application facilitated oil splitting by 65–75% as compared to the control [6].

This paper presents data evaluating application efficiency of the first national microbial preparation Rhodobel designed for disposal of oil and oil-processing products in soil.

Objects and Methods

The following investigation objects were used in subsequent experiments:

- microbial preparation Rhodobel composed of dry biomass of strains-degraders *Rhodococcus erythropolis 5D, Rhodococcus opacus 7F, Rhodococcus opacus 29 F, Rhodococcus rubber 1B*
- oil (d=0.85 g/cm^3, C–85.6%, H–12.7%, O+S–1.7%)
- diesel oil (d=0.82 g/cm^3)
- peat characterized by degradation degree R=20–25%.

Examination of Rhodobel ability to intensify decomposition of crude oil and diesel fuel was carried out in glass 750 ml vessels containing 100 g of sod-podzol soil distinguished by the following agrochemical parameters: pH_{KCl} 6.3; P_2O_5 – 19.2 mg/700g soil; K_2O – 15.8 mg/700 g soil; base saturation degree – 68%, humus – 1.6%, total nitrogen – 0.15%. Soil humidity was maintained at the level 60% of overall moisture capacity.

Soil polluted with oil (0.5% and 1% concentration) and diesel fuel at 1% concentration was treated with microbial preparation at the ratio of oil or diesel oil to Rhodobel 1:0.025. Peat (1% of soil mass) was used as a support for strains-degraders, for oil sorption and as a source of humus.

Tests of microbial preparation Rhodobel to break down oil and diesel fuel in field experiments were performed at Polotsk branch of Druzhba pipeline sized 40x15 m. Sod podzol soil (pH_{KCl} 7.35; P_2O_5 – 22.9 mg/700g soil; K_2O – 12.7 mg/700 g soil) was artificially contaminated with oil at 0.5% and 1% concentrations and with diesel fuel at 1% concentrations. The technology of soil bioremediation using Rhodobel preparation comprised the following stages: preparation of microbial suspension from dry bacterial mass – active principle of Rhodobel; application of peat on contaminated soil surface; application of microbial suspension onto polluted soil at the ratio oil or diesel fuel : Rhodobel 1:0.025.

Microorganisms capable to consume oil and diesel fuel as the sole carbon source (both indigenous and introduced) were counted on agar mineral medium of the following composition, g/l: NaCl – 0.5, $MgSO_4 \cdot 7H_2O$ – 0.8, KH_2PO_4 – 0.7, $(NH_4)_2HPO_4$ – 1.5, carbon source (oil or diesel fuel) – 1.0. Identification of introduced rhodococcal strains was performed by indirect immunofluorescence technique [7].

Quantitative determination of aliphatic hydrocarbons was conducted at IR spectrometer Specord M80 with original PC software. Mono- and polycyclic aromatic hydrocarbons were assessed by UV-spectrophotometry at Specord M40 unit.

RESULTS AND DISCUSSION

Control samples of soil used in the lab model experiment comprised microorganisms able to digest oil as the sole carbon source in concentration $0.1–2.7 \cdot 10^7$ cells per 1 g absolutely dry soil. Oil introduced into soil stimulates development of strains-degraders. Seven days after application of oil at 0.5% concentration the titer of indigenous degradative microorganisms rose from $0.3 \cdot 10^5$ to $1.9 \cdot 10^8$ cells/1g absolutely dry soil and maintained at this level throughout the whole experiment. Indigenous microbial cultures stimulate degradation of aliphatic oil constituents after 28 days by 50% and aromatic as more persistent compounds – by 20% (Table 1,2). Introduction of degrading microorganisms into soil facilitates oil disruption. Oil aliphatic components were broken down using Rhodobel preparation by 70.6%, aromatic components – by 33.4%. Rhodococcal population administered into soil as a part of microbial preparation was represented in 1g of absolutely dry soil of original sample by cell number $1.5 \cdot 10^8$. This value was kept up in soil approximately at the initial level and even tended to rise by 21 days gradually declining by day 28, correlated with reduction of oil levels in soil. Oil at 1% concentration also stimulated development of indigenous microbial population in contaminated soil samples. Population density increased from $0.1 \cdot 10^7$ to $1 \cdot 10^7$ cells/g absolutely dry soil after 7 days and later on stabilized at the level $1.2 \cdot 10^7$ cells/g absolutely dry soil. Oil at 1% concentration was decayed by 34.4% (aliphatic component) and by 5.6% (aromatic component) with active involvement of indigenous microbiota. Microbial preparation accelerates oil degradation process. It contributes into decomposition of aliphatic oil constituents by 50%, and aromatic components by 11.1% in 25 days. 7 days later a slight drop in population density of rhodococci was registered (from $2 \cdot 10^8$ to $8 \cdot 10^7$ cells/g absolutely dry soil) caused by adaptation to elevated oil levels and competition with indigenous species for nutrient source. In the next 28 days, the population was decreased due to reduced oil content in soil and hence, limitation of available carbon sources.

Table 1 Degradation of aliphatic oil components in soil under conditions of laboratory model

Variants	Hydrocarbon residual amount, %				Degradation efficiency, %
	7 days	14 days	21 days	28 days	
Soil+0,5% oil	86,7	86,7	86,7	50,0	50,0
Soil +0,5% oil + Rhodobel	82,4	82,4	70,6	29,4	70,6
Soil +1,0% oil	93,7	90,6	90,6	65,6	34,4
Soil +1,0% oil + Rhodobel	85,3	85,3	85,3	50,0	50,0
Soil +1,0% diesel oil	97,2	88,9	80,6	63,9	36,1
Soil +1,0% diesel oil + Rhodobel	73,3	68,9	62,2	46,7	53,3

Table 2 Degradation of aromatic oil components in soil under conditions of laboratory model

Variants	Hydrocarbon residual amount, %				Degradation efficiency, %
	7 days	14 days	21 days	28 days	
Soil+0,5% oil	98,0	87,0	80,0	80,0	20,0
Soil +0,5% oil + Rhodobel	96,7	70,8	68,3	66,6	33,4
Soil +1,0% oil	98,3	98,3	98,3	94,4	5,6
Soil +1,0% oil + Rhodobel	98,9	96,1	94,4	88,9	11,1

Table 3 Degradation of aliphatic oil components in soil in field trials

Variants	Hydrocarbon residual amount, %						Degradation efficiency, %
	28 days	**37 days**	**61 days**	**75 days**	**88 days**	**110 days**	
Soil +0,5% oil + Rhodobel	75,0	50,0	25,0	20,0	7,5	0	100
Soil +1,0% oil	66,7	66,7	66,7	53,3	46,7	33,3	66,7
Soil +1,0% oil + Rhodobel	25,0	15,0	10,0	3,0	1,5	0	100
Soil +1,0% diesel oil + Rhodobel	50,0	10,0	10,0	5,5	0,2	0	100

Table 4 Degradation of aromatic oil components in soil in field trials

Variants	Hydrocarbon residual amount, %						Degradation efficiency, %
	28 days	**37 days**	**61 days**	**75 days**	**88 days**	**110 days**	
Soil +0,5% oil + Rhodobel	25,0	18,8	17,5	15,0	12,5	5,0	95,0
Soil +1,0% oil	66,7	66,7	63,3	56,7	50,0	33,3	66,7
Soil +1,0% oil + Rhodobel	50,0	15,0	5,0	5,0	5,0	5,0	95,0

Diesel fuel, like oil, stimulates development of indigenous microflora in soil. In the first 7 days after diesel oil contamination the population of indigenous species in soil was increased by 100 times. In subsequent period it remained virtually stable, achieving 36% degradation of diesel fuel by 28 days. Introduction of degradative microorganisms into polluted soil intensified decomposition rate. For comparison, after 28 days of monitoring diesel fuel was degraded by 53.3%. In the course of observation period population of species degrading diesel oil ranged from $1.2 \cdot 10^8$ to $8.1 \cdot 10^8$ cells/g absolutely dry soil. Introduced Rhodococcal population rose by 14 days from $1.9 \cdot 10^8$ to $6.2 \cdot 10^8$ cells and declined by 28 days to $1.2 \cdot 10^7$ cells/g absolutely dry soil, caused by shortage of diesel oil substrate in later period.

Efficiency of Rhodobel in soil remediation proved in lab conditions was successfully corroborated in industrial trials with soil polluted with oil (0.5% and 1.0%) and diesel fuel (1.0%). Diesel fuel (1.0%) in soil containing microbial preparation was decomposed by 99.98% in 3 months (Table 3). Diesel fuel reduces population of degradative organisms in soil. Decrease of microbial mass in polluted soil was registered during 2 months resulting in 90% breakdown of the substrate. Later degradative population was rising, reaching steady state at $2.5 \cdot 10^7$ cells/g abs.dry soil. Introduction of degradative species into diesel oil-contaminated soil as a part of Rhodobel preparation led to a fall in population density from $3.3 \cdot 10^7$ to $0.8 \cdot 10^7$ cells/g abs.dry soil during the first month of observations. This may be explained by competition for nutrients with indigenous microflora. Peak of introduced population density was registered after 37 days, with resultant degree of diesel oil decomposition reaching 90%. Rhodobel application promoted more efficient oil degradation in soil under industrial trial conditions. Aliphatic oil components (0.5%) were completely decomposed by 3.5 months, whereas degree of degradation for recalcitrant aromatic fraction reached 95% (Table 4). Rhodococcal population introduced into contaminated soil as an active principle of Rhodobel decreased within 30 days from $3.0 \cdot 10^7$ to $0.3 \cdot 10^7$ cells/g abs. dry soil, caused presumably by competitive relations with indigenous microflora. The peak of introduced rhodococcal population density was registered by 2 months of industrial trials. Corresponding figures for degradation of aliphatic and aromatic oil constituents equaled 75.0 and 82.5%. Further on population of introduced microbial species was stabilized at the level $0.4 \cdot 10^7$ cells/ g abs. dry soil. Total cell number of degradative microorganisms in soil polluted with 0.5% oil concentration was gradually falling by the end of trials, correlated with oil degradation. Aliphatic oil components (1.0%) in soil treated with Rhodobel were decayed by 100% in 3.5 months, whereas aromatic components were degraded by 95%. in control soil sample (without Rhodobel) aliphatic and aromatic oil constituents were degraded by 66.7% of initial levels. Population of introduced microbial species in soil contaminated with oil (1%) declined after the first month of observations from $4.1 \cdot 10^7$ to $0.4 \cdot 10^7$ cells/1 g abs. dry soil. By 2 months rhodococcal population in soil attained $3 \cdot 10^7$ cells/1 g abs. dry soil and degradation degrees reached 90.0% and 82.5% for aliphatic and aromatic components, respectively. Oil decay in soil was accompanied by gradual decrease of rhodococcal population down to $0.1 \cdot 10^7$ cells/1 g abs. dry soil.

Rhodobel application guarantees disposal of oil pollution in soil within 5 months during the spring-summer season, simple remediation procedures, and complete ecological safety.

REFERENCES

1. L.A. Golovleva: *Biotechnologia,* (5), **60** (1992).
2. V.p. holodenko et.al.: *Abstr. Conf. Introduction of Microorganisms into Environmental Media.* Moscow, **34** (1994). (in Russian).
3. R.K. AndrEson et.al. : *Abstr. Conf. Introduction of Microorganisms into Environmental Media.* Moscow, **10** (1994). (in Russian).
4. I.A. Borzenkov et.al : *Abstr. Conf. Introduction of Microorganisms into Environmental Media.* Moscow, **14** (1994). (in Russian).
5. T.V. Koronelli: *Abstr. Conf. Introduction of Microorganisms into Environmental Media.* Moscow, **53** (1994). (in Russian).
6. Z.I. Finkelstein et.al : Abstr. Conf. Biotechnology of Environmental Protection, *Pushchino*, **5** (1994). (in Russian).
7. P.A. Kozhevin: *Natural microbial populations*, MSU Publ, Moscow, 1989. 175 p. (in Russian).

Chapter 16

TRANSFORMATION OF L-CYSTINE AND L-CYSTEINE INTO TAURINE WITH ENZYMATIC SYSTEMS OF BOVINE LIVER CELLS

Anna V. Soboleva, Alla A. Krasnoshtanova, and Igor A. Krylov
Department of Biotechnology, D.I. Mendeleyev University of Chemical Technology
Moscow, Russia

ABSTRACT

The kinetics of transformation of the sulfur amino acids L-cystine and L-cysteine into taurine with enzymatic systems of bovine liver cells has been investigated. This stage of the technological process has been investigated and the mathematical model of the process has been created. The transformation of L-cystine and L-cysteine is shown to conform to the scheme of Michaelis-Menten series-parallel transformations taking into account of inhibition by the end product and its inactivation. The optimum conditions for realization of transformation have been determined: 40°C; pH 1.5 and 3.0, respectively, and 5-fold loading of the enzyme by equal shares every two hours of the process, providing 80-85% yield of taurine by substrate weight.

Key words: L-cysteine, L-cystine, taurine, sulfur amino acids

INTRODUCTION

Taurine is a product of metabolism of the sulfur amino acids L-cystine and L-cysteine. It plays an important role in providing various human functions connected to eyesight, aggregation of thrombocytes, and an insulin activity etc. It is used as an individual preparation to enhance the functions above-stated in medicine. Also, it is important for fodder and food-processing industry [1]. Taurine is a substance that can be synthesized chemically quite easy and also it can be obtained by transformation of the above-mentioned amino acids as either an individual compound or a part of protein hydrolyzates under action of a complex of bovine liver cell enzymes [2], which can be considered as an enzymatic preparation (EP).

In the latter case it is possible to use keratin containing raw materials, which are rich in L-cystine and are the waste of meat-processing industry [3]. Such enterprises can be considered as a source of enzymes for realization of the transformation involved, the enzymes being enzymatic systems of cattle or swine liver cells and the substrate not being used high-priced individual L-cystine and L-cysteine but hydrolyzates or enzymatic hydrolyzates of L-cystine and L-cysteine containing proteins on the base of keratins. In given system L-cystine (L-cysteine) – liver cells a rather complicated scheme of series-parallel enzymatic transformations (Fig.1) that would provide a technologically reasonable output of taurine should be realized.

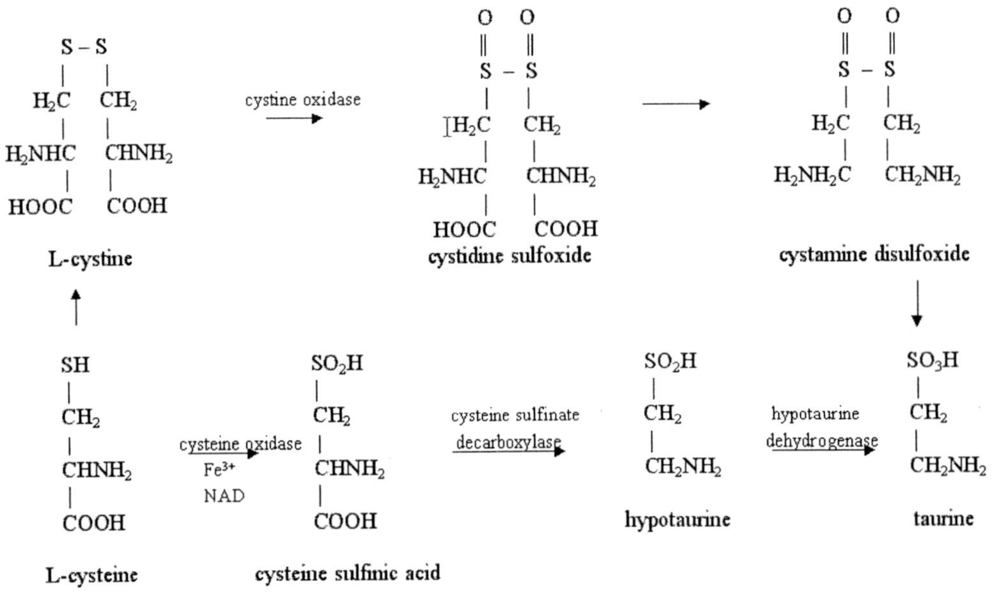

Figure 1. The pathway from L-cystine and L-cysteine into taurine with enzymatic systems of bovine liver cells

The Purpose of the Investigation

Taking into account comparatively large number of enzymes participating in the metabolism of the above-mentioned amino acids into taurine, and also a complex structure of hydrolyzates of keratin containing raw materials a model system L-cystine, L-cysteine – cattle liver cells was chosen as an object of research at the initial stage. On this model example the basic regularities and mechanism of transformation of individual amino acids into taurine have been investigated.

Materials and Methods

L-cystine and L-cysteine consisting not less than 98% of the base material were purchased from Merck. Cattle liver cells treated by the technique [4] or [5] have the activity of not less 2000 un/g of the dried weight.

Other reagents which provided realization of analyses and titrants were prepared with distilled water with application of reactants of higher quality.

Transformation of L-cystine and L-cysteine into taurine was carried out in a glass reactor of the volume of up to100 ml provided with a mixer and several necks for sampling and adding some titrant and also supplied with the electric drive, the thermometer, a water jacket, and electrodes of pH-meter. The constant temperature in the reactor was kept accurate to ±0.5°C by water supply from the thermostat into the jacket of the reactor. Before each experiment a fresh water solution of an amino acid of given concentration was prepared and added into the reactor preheated up to the required temperature. The required temperature having been reached, the pH of the solution was adjusted by adding some titrant in the reactor (0.01N water solution of hydrochloric acid or sodium hydroxide). Thereafter special pretreated liver cells were added in the reactor. The moment of their addition was accepted as the time of the beginning of transformation. In the course of the process sampling was carried out every other definite period of time, the samples being cooled to the temperature 4-6°C in the ice bath. In the samples obtained the amount of the unreacted substrate and taurine formed were determined by the techniques [6] and [7], respectively.

During investigation of the quantitative regularities of transformation the effect on the rate of the reaction by the following parameters: temperature, pH, initial concentrations of liver cells and a substrate within the ranges: 30-60°C, 1.5 -7.0, 1.3 – 3.2 g of dried weight/liter, and 10 -100 g/liter has been studied.

Evaluation of the experimental data was carried out with the use of standard programs of integration of the differential equations and optimization of their parameters followed by comparison of the results of calculation with the experimental data according to Fisher's test, being used for check of adequacy of the developed mathematical model.

Results and Discussion

In the course of experimental researches the series of kinetic curves of substrate consumption and taurine accumulation, which corresponded to changing one of the parameters, were obtained at the fixed values of others.

The result of such investigations has shown the following: irrespective of conditions of transformation the reacted substrate was transformed into taurine practically completely (not less than 95%), that is illustrated in figure 2.

The regularity observed and absence of points of inflection on taurine accumulation plots have allowed us to consider the transformation given in figure 1 as series of rapid enzymatic reactions, where a taurine forming stage is a limiting one. This has permitted to investigate the kinetics of taurine formation from the above-mentioned amino acids as a one-stage enzymatic reaction complicated with the presence of inhibition.

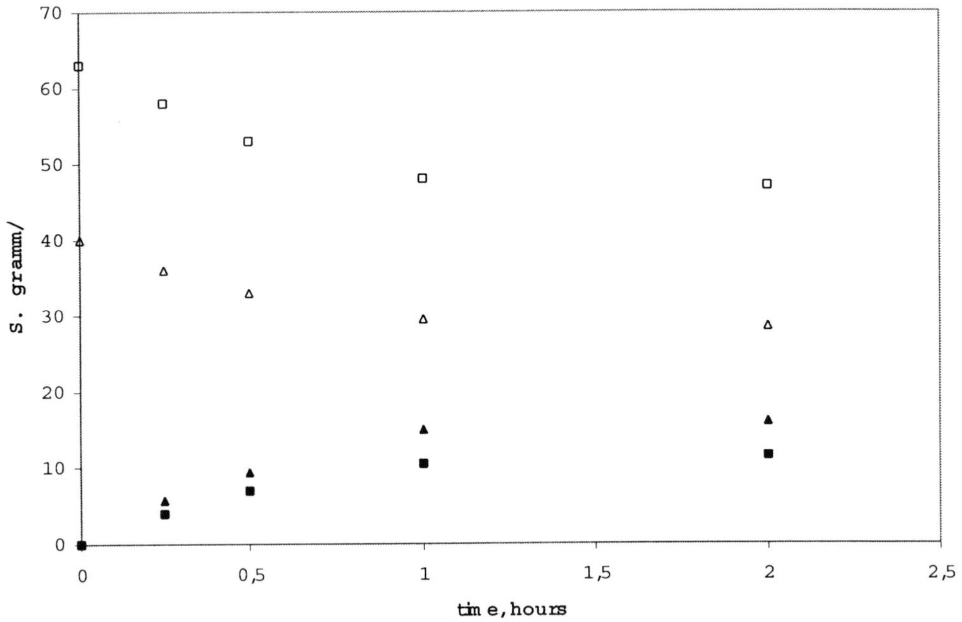

Figure 2. L-cystine and L-cysteine consumption and taurine accumulation at transformation these subsrates into taurine. The condition: temperature – 40 °C, pH 2.0 – 2.5, the initial activity of enzymatic systems of bovine liver cells – 8 units/ml

The first stage of investigation included the study of relationship of the initial rate of every amino acid transformation from the concentration of liver cells or liver cell enzymatic activity at fixed values of other parameters of the process. It was shown that they can be considered as linear in the EP concentration range within 0 – 1.5 g/liter, the correlation coefficient (r_c) being not less than 0.96.

In the following series of functions of the initial rate of transformation vs. the initial substrate concentration in Lineweaver-Burk double-reciprocal plot and from the temperature in Arrhenius plot were studied. On Lineweaver-Burk plot (Figure 3) and on Arrhenius plot the straight lines were observed, that have allowed us to calculate the Michaelis-Menten constant (K_M) and the free energy of activation of transformation for each substrate into taurine.

Linear functions in figure 3 allowed us to apply the Michaelis-Menten rate equation of enzymatic transformations for analyzing of initial parts of kinetic curves:

$$S + E \underset{}{\overset{K_M}{\rightleftarrows}} [I] \xrightarrow{k_2} P + E \qquad (1),$$

where S, E and P represent the substrate concentration, the concentration of the free enzyme, and the product concentration, respectively, and [I] stands for the reversibly formed enzyme-substrate complex, which is decomposed to form taurine and the free enzyme.

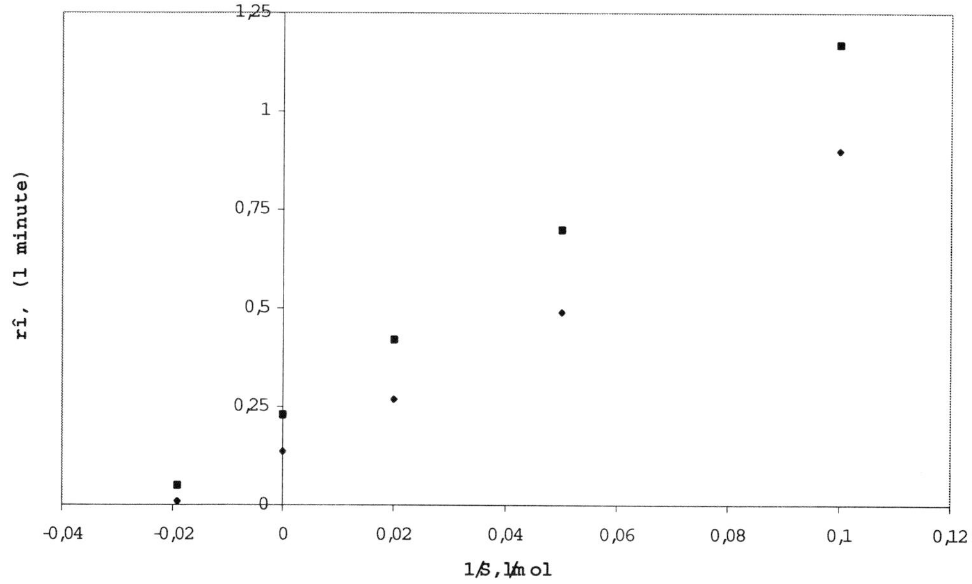

Figure 3. Relationship of the initial rate of transformation of L-cystine and L-cysteine into taurine vs. the initial substrate concentration in Lineweaver-Burk double-reciprocal plot. The condition: temperature – 40 °C, pH 2.0, the initial activity of enzymatic systems of bovine liver cells – 8 units/ml

However, the equation (1) does not take into account the inhibition revealed. Character of enzyme inhibition can be rather complex and include inhibition by the end product (taurine), and also free enzyme inactivation or inactivation of the enzyme as constituent of enzyme-substrate complex.

To solve this problem we carried out independent experiments on transformation of substrates investigated into taurine at various taurine concentrations ranged within (1.6 – 6.4) $\cdot 10^{-3}$ mol/liter.

The results of these experiments have shown that the inhibition of transformation by the end product takes place. When the initial taurine concentration increases, the initial velocity of the process for each substrate decreases steadily. Simple processing the given series of experiments on the initial rates represented in plots of $1/v_0$ vs. C_P has given linear dependences on each substrate (Figure 4), that has permitted us to add the equation of reversible free enzyme inhibition into the transformation pathway (2) with the constant K_I and estimate its value.

$$P + E \underset{}{\overset{K_I}{\rightleftarrows}} [II] \qquad (2),$$

where [II] – an inactive intermediate complex.

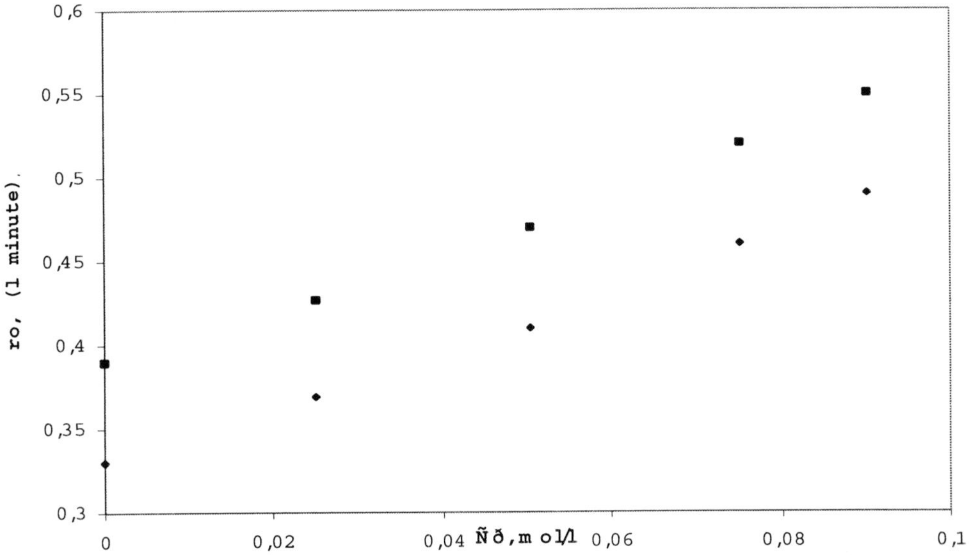

Figure 4. Relationship of the initial rate of transformation of L-cystine and L-cysteine into taurine vs. the initial content in parent solution. The condition: temperature – 40 °C, pH 2.0, the initial activity of enzymatic systems of bovine liver cells – 8 units/ml

On the basis of results of the preliminary processing of the experimental data it is possible to integrate the system of differential equations, which corresponds to the united transformation equations (1) and (2). However, this system is not able to describe exactly a course of kinetic curves in the interval of changing parameters a priori, as in the experiment (Figure 2) the transformation of amino acids does not proceed to the complete exhaustion of a substrate, but it stops at some level, that indicates the presence of additional inhibition factors.

In this case it is possible to assume, that a fact of inhibition is caused by inactivation of the enzyme that is responsible for the transformation of enzyme either in free form or in a structure of enzyme-substrate complex. Such transformations can be presented as the following equations:

$$E \xrightarrow{k_{I1}} E_{inactive} \text{ and (or)}$$
$$[I] \xrightarrow{k_{I2}} [I]_{inactive} \qquad (3)$$

where E inact and (or) [I] inact represent an inactive form of free enzyme and the enzyme in a structure of the enzyme-substrate complex.

Therefore, the integral processing of the experimental data was carried out with addition of equation (3) into equations (1) and (2) for series of kinetic curves at various initial substrate concentrations and temperatures.

In the course of integral processing for every kinetic curve the values of the appropriate constants k_I in equations (3) were selected, achieving the adequacy to experimental data. As a result, it was possible to calculate the value of the constant k_I that remained practically inalterable in the series of kinetic curves, where only the initial substrate concentration was changed. And for a series of the curves obtained at various temperatures, its value accrued in a cymbate way with increase of temperature. In the latter case the linear dependences were obtained for each of substrates that allowed us to estimate the value of the energy of activation (E) of transformation of each amino acid into taurine on the basis of established values of the enzyme inactivation constant in Arrhenius axes.

At completion phase of the researches the optimization of the values of parameters of equations (1) - (3): K_M, K_I, k_2, and E, describing the kinetics of the process researched was carried out. The values of constants and energies of activation are given below in the table.

The established values of constants k_2 in series of kinetic curves corresponding to various values of pH were used for the description of relationships, which are typical for enzymatic reactions and given in figure 5, and which should obey the equation:

$$K_{2\,ef} = k_2^* / (1 + [H^+] / Ka) \quad (4),$$

that establishes the relationship of enzyme transformation velocity and concentration of hydrogen ions in a solution. In the plot of k_{ef}^{-1} vs.$[H^+]$ the linear dependences as the designed factors of correlation was not less, than 0.95 should be observed. The linear kind of the given function has allowed us to estimate numerical values of constants k_2 and Ka, which further were specified during optimization of given series of kinetic curves, and they are submitted in the table.

The regularities established above have allowed us to find optimum conditions of realization of process of transformation of amino acids into taurine. The results of the calculations confirmed with the experiments have shown that the optimum conditions for transformation of 0.33 M L-cystine and L-cysteine into taurine at ratio of a substrate to liver cells 0.16 -2 are the temperature of 40 °C, pH 3.0.Thus the output of the product required appears to be low and amount 20% and 30%, respectively. The calculations have shown that such a low yield of taurine is caused by decrease of the enzymatic system activity by the end product inhibition of transformation, but mostly (more than 70 %) according to equation (3).

To find out the reason of the researched enzymatic liver sell system inactivation submitting to the first order equation we have assumed, that increased temperature of the transformation was the main one. The hypothesis given was checked up in two series of experiments, when the suspension of prepared liver cells was kept with agitation at various temperatures and time, then transferred in a substrate solution, and investigated taurine accumulation kinetics. The results of these experiments have allowed us to draw an unequivocal conclusion that the liver cell active protein thermo inactivation effect is present.

To achieve a technologically acceptable output of a base product (taurine) it was of interest to find a certain optimum mode of realization of L-cystine and L-cysteine transformation at most possible concentrations of them in hydrolyzates of keratin containing raw materials. If such a hydrolyzate is prepared on the basis of 10 % keratin suspension, the most possible of L-cystine (L-cysteine) concentration will not exceed 0.06mol/liter.

Table 1. The effective constants and activation energies of transformation of L-cystine and L-cysteine into taurine with enzymatic systems of bovine liver cells

Substrate	Stage of substrate transformation				Stage of inactivation			
	Км mol/l	k^*_2 1/s	Ка mol/l	energy of activation (E), кJ/mol	Ки mol/l	k^*_2 1/s	Ка mol/l	energy of activation (E), кJ/mol
L-cystine	1268,21 ± 101,53	6,02 ± 0,24	48,42 ± 2,91	62,41 ± 2,53	2,13 ± 0,13	$(1,5 \pm 0,01) \cdot 10^{-3}$	$0,19 \pm 1,32 \cdot 10^{-3}$	6,7 ± 0,335
L-cysteine	76,93 ± 3,82	$0,03 \pm 2 \cdot 10^{-3}$	$1,49 \cdot 10^{-3} \pm 0,07 \cdot 10^{-3}$	14,12 ± 0,63	2,13 ± 0,13	$(3,1 \pm 0,02) \cdot 10^{-3}$	$5,29 \cdot 10^{-4} \pm 3 \cdot 10^{-6}$	20,14 ± 1,21

*) – the value of the constant independent of the pH medium value at the temperature of 40°C

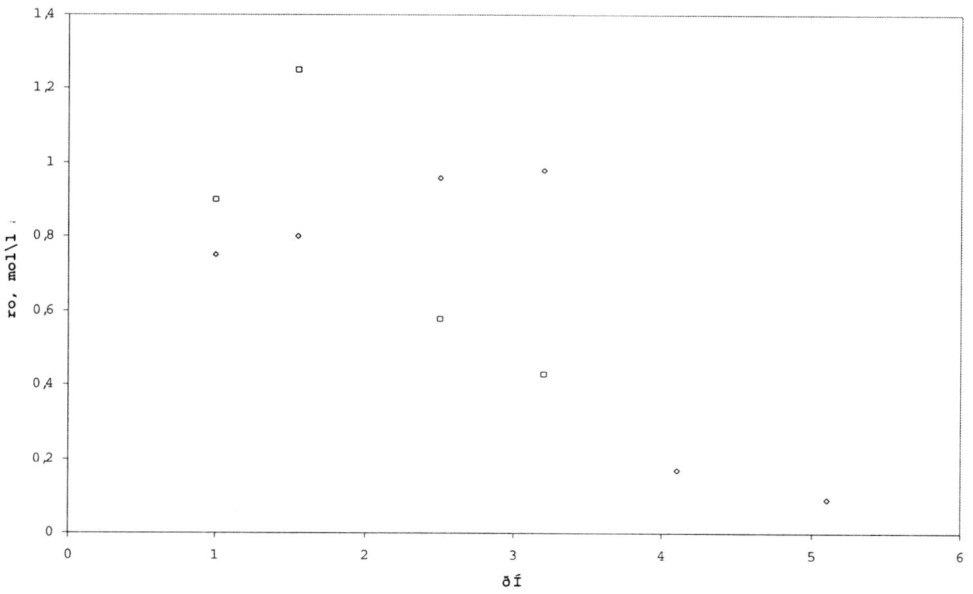

Figure 5. Relationship of the initial rate vs. pH of solution at transformation of L-cystine and L-cysteine into taurine. The condition: temperature – 40 °C, the initial substrate concentration - 0.33 mol/l, the initial activity of enzymatic systems of bovine liver cells – 8 units/ml

Whereas the significant part of enzyme liver cell preparation is inactivated according to the first order equation because of thermal denaturation of its active proteins, to increase the yield of taurine from the above-mentioned amino acids is possible by fractional loading of the enzyme during the time of transformation.

Based on a condition of 95% inactivation of the enzymatic system the time intervals of fractional enzyme preparation loading and its amount were determined. In addition, the purpose was to minimize a norm of enzyme preparation rate and to reach technically acceptable output of taurina, which was solved by the appropriate calculations, and the result obtained was check up experimentally.

The results of calculation have shown, that process of L-cystine and L-cysteine transformation into taurine is expedient for carrying out at the temperature of 40 °C, pH 1.5 and 3.0, respectively, with 5-fold loading of the enzyme preparation by equal shares every other two hours. Under these conditions the above-mentioned amino acids should be transformed into taurine up to 80-85% during the full time of transformation not more than 10 hours, the total activity being 1600 units/mmol of substrate.

The results of calculations were checked up experimentally. They are demonstrated in figure 6, which shows that the taurine accumulation curve grows steadily, and the experimental data confirm the end product value calculated in full.

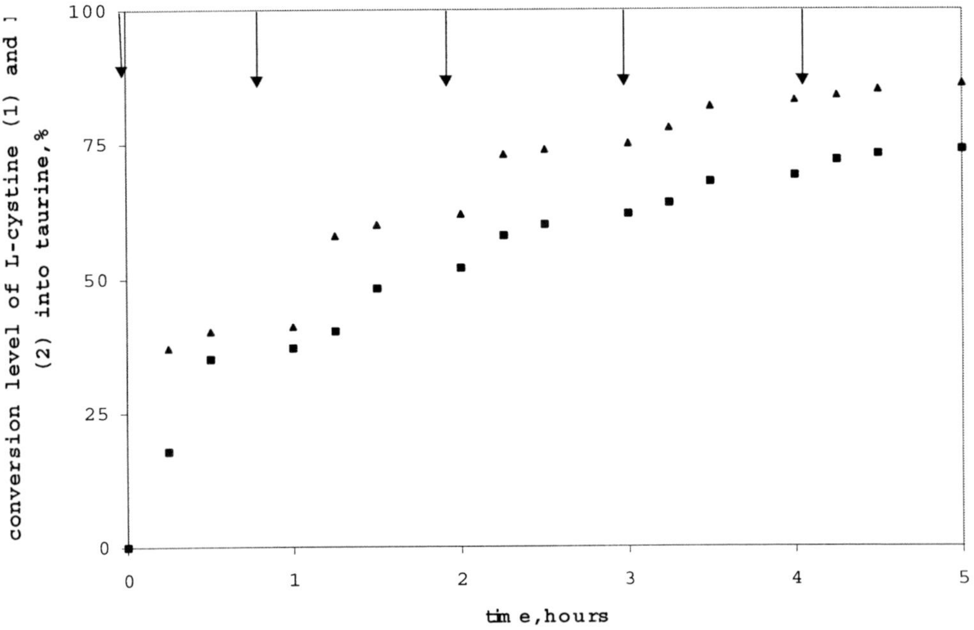

Figure 6. Relationship of conversion level of L-cystine (1) and L-cysteine (2) into taurine at transformation of amino acids with enzymatic systems of bovine liver cells. The condition: temperature – 40 °C, pH 1.5 (1) and pH 3.0 (2), total activity of enzymatic systems of bovine liver cells – 16 units/ml Designation: ↓ - load time of EP

References

1. V.S. Gurevich: *Taurine and Excitable Cell Function*, Moscow, 1986 (in Russian).
2. L.I. Nefedov: *Taurine (Biochemistry, Pharmacology, and Medical Application)*, Grodno, 1999 (in Russian).
3. A.V. Bertudina, A.V. Neckludova, A.N. Ivankin, B.S. Carpo, and A.V. Ocoka: Investigation of Keratin Containing Raw Material Acid Hydrolysis. *II All-Russian Conf. Progressive environmentally safe technology for storage and processing of agricultural products for creation of high nutritional and biological valuable food products*, Uglich, 1996 (in Russian).
4. A.D. Neckludov, V.V. Tcibonov: *Antibiotics and Med. Biotechnology*, 1985 – v.30, №3. 176-179 p. (in Russian).
5. A.D. Neckludov: *Chem. Pharm. J.*, 1985, №8, 998-1001p. (in Russian).
6. L.A. Osterman: *Chromatography of Proteins and Nucleic Acids*, Moscow, 1985 (in Russian).
7. F. Kritchfield: *Analysis of Basic Functional Groups in Organic Compounds*, Moscow, 1965 (in Russian).

SECTION 2.
BIOGEOTECHNOLOGY

Chapter 17

Variability of Chemolithotrophic Microorganisms as a Basis for the Regulation of Their Activity under Extreme Conditions

T.F. Kondrat'eva [1]***, O.V. Tupikina*** [1]***, S.N. Ageeva*** [1]***,***
V.D. Samorukova [2] ***and G.I. Karavaiko*** [1]

[1]Institute of Microbiology, Russian Academy of Sciences, Moscow, 117811 Russia

[2]Institute of Complex Utilization of Mineral Raw Materials and Wastes, Russian Academy of Natural Sciences, Lyubertsy, Russia

Abstract

Pulsed-field gel electrophoresis was used to compare the chromosomal DNA restriction profiles from the strains of acidophilic chemolithotrophs *Acidithiobacillus ferrooxidans, A. thiooxidans, Sulfobacillus sibiricus* and *Ferroplasma acidiphilum* isolated from different environments. The DNA restriction patterns were specific for each strain. The adaptation of strains to new oxidation substrates or to increased concentrations of heavy metal ions (zinc) or to toxic elements (arsenic), caused either the changes in the number and size of the DNA fragments generated by the restriction endonucleases or amplification of certain fragments, or, in the case of *A. ferrooxidans* strains, the changes in the plasmid profiles. The genotypic and phenotypic characteristics of *A. ferrooxidans* strains were affected not only by the chemical composition of the oxidation substrate, but also by the electrophysical and crystallochemical peculiarities of one and the same mineral substrate. All *A. ferrooxidans* strains differed in the growth rate and oxidation activity on lowering the pH value from 1.1 to 0.9 within three employed temperature regimes (4-6°C, 20-22°C and 28-30°C). Variability of acidophilic chemolithotrophs, causing their strain polymorphism, is considered to be the result of adaptation to changing environmental conditions, primarily the oxidation substrate.

Keywords: acidophilic chemolithotrophs, chromosomal DNA, plasmid profiles, substrate, electrophysical properties, variability

Variability reflects the response of organisms to changes in environmental conditions and is realized either through nonheritable phenotypic variations within the normal range of genotypic response, or through inheritable genotypic changes resulting in the formation of the organism with new properties.

Variability of chemolithotrophic microorganisms as a basis for the regulation of their development under extreme conditions is considered in this paper by the example of acidophilic chemolithotrophs. This is a relatively small group of cultivated, phylogenetically unrelated microorganisms including mesophilic and moderately thermophilic Gram-negative bacteria of the genera *Acidithiobacillus (A. ferrooxidans, A. thiooxidans, A. caldus)* and *Leptospirillum (L. ferrooxidans)*, mesophilic and moderately thermophilic Gram-positive bacteria of the genus *Sulfobacillus (S. sibiricus, S. thermosulfidooxidans, S. acidophilus)* and mesophilic and thermophilic archaea of the genera *Ferroplasma (F. acidiphilum), Acidianus (A. brierleyi), Metallosphaera (M. sedula)* and *Sulfolobus (S. metallicus).* The common property of all these microorganisms is their ability to derive energy at low pH values from the oxidation of inorganic substrates, such as ferrous iron, elemental sulfur, its reduced compounds, and sulfide minerals. Therefore, acidophilic chemolithotrophs are isolated from rocks, sulfide ores, biohydrometallurgical technological pulps and other extreme econiches characterized by low pH and containing the above mentioned oxidation substrates. Acidophilic chemolithotrophs develop under varying environmental conditions: permanently changing oxidation substrates, increasing concentrations of heavy metal ions and toxic elements in the liquid phase during the oxidation of sulfide minerals, lowering of pH to values much below the optimal range with the oxidation of elemental sulfur and its reduced compounds, changes in the temperature regime due to self-heating of the pulp in the course of sulfide mineral oxidation. In order not only to survive but to successfully develop, acidophilic chemolithotrophs should be able to adapt to novel environmental conditions and possess labile mechanisms for genome regulation providing their variability.

The aim of the present work was to study genotypic and phenotypic variability of acidophilic chemolithotrophs upon switching their metabolism to oxidation of a different substrate, lowering pH value, changing the temperature regime, or increasing the content of heavy metal ions or toxic elements.

Materials and Methods

Bacterial Strains and Growth Conditions

Experiments were conducted with 17 strains of *A. ferrooxidans,* 3 strains of *A. thiooxidans*, 3 strains of *S. sibiricus* and 4 strains of *F.acidiphilum* isolated from different habitats and kept in the culture collection of the Laboratory of Chemolithotrophic Microorganisms, Institute of Microbiology, Russian Academy of Sciences.

Batch cultivation was performed under the conditions optimal for each strain [1-3]. One of the following substrates was used as the energy source: ferrous iron, elemental sulfur, pyrites of different origin and gold-containing pyrite-arsenopyrite ore concentrates from different deposits.

Preparation of Intact Chromosomal DNA and its Restriction Analysis

Harvesting and washing of biomass, preparation of native genomic DNA and its restriction analysis were performed as described earlier [1-3].

Isolation of Plasmid DNA and its Analysis

Isolation of plasmid DNA from *A. ferrooxidans* cells and analysis of plasmid profiles by electrophoresis were performed as described earlier [4].

Physicochemical Properties of Pyrites

Physical and electrophysical properties of the pyrites, and their crystallochemical peculiarities were analyzed by the conventional methods: density – by the microburet method; microhardness – with a PMT-3 microhardness gage (Russia) at 100-g load and 10-s exposure; crystallochemical peculiarities – with a microprobe analyzer of a scanning electron microscope JSM-5300 (Japan); the conductivity type and the values of resistance and thermoelectric power (TEP) were analized at constant temperatures of "hot" electrode.

Results and Discussion

Figure 1 shows strain-specific polymorphism of the chromosomal DNA structure analyzed by pulsed-field gel electrophoresis in several acidophilic chemolithotrophs. Each strain appeared to have a unique restriction pattern of chromosomal DNA. This genotypic characteristic is stable enough to be used as a reliable feature for the identification of new strains, for the monitoring of known strains in their natural habitats and in technological processes, and also in studies of spontaneous and induced variability in natural environment and in the experiments. The adaptation of acidophilic chemolithotrophs to new oxidation substrates was often accompanied by the changes in the chromosomal DNA structure: changes in the number and size of certain restriction fragments, in the number of generated fragments of the same size, or the amplification of particular fragments was revealed (Figs. 2 and 3). It is notable that the adaptation of strain TFBk to one and the same mineral (pyrite) but obtained from different deposits, either caused changes in the chromosomal DNA structure, or changes did not occur. The results obtained allow us to suggest that the genotypic variability of strains is affected not only by the chemical composition of the oxidation substrate, but also by the physico-chemical, electrophysical properties and crystallochemical peculiarities of one and the same mineral substrate.Earlier, Grudev [5] and Karavaiko et al. [6] reported that one and the same strain oxidized different pyrites with different activity. However, this observation did not receive further development. In this work, we studied the influence of physico-chemical peculiarities of two types of pyrites on the phenotypic and genotypic characteristics of two *A. ferrooxidans* strains, TFBk and TFV-1. Strain TFV-1 was isolated from a relatively simple in mineralogical composition poor copper ore, and strain TFBk – from a complex gold-bearing pyrite-arsenopyrite ore.

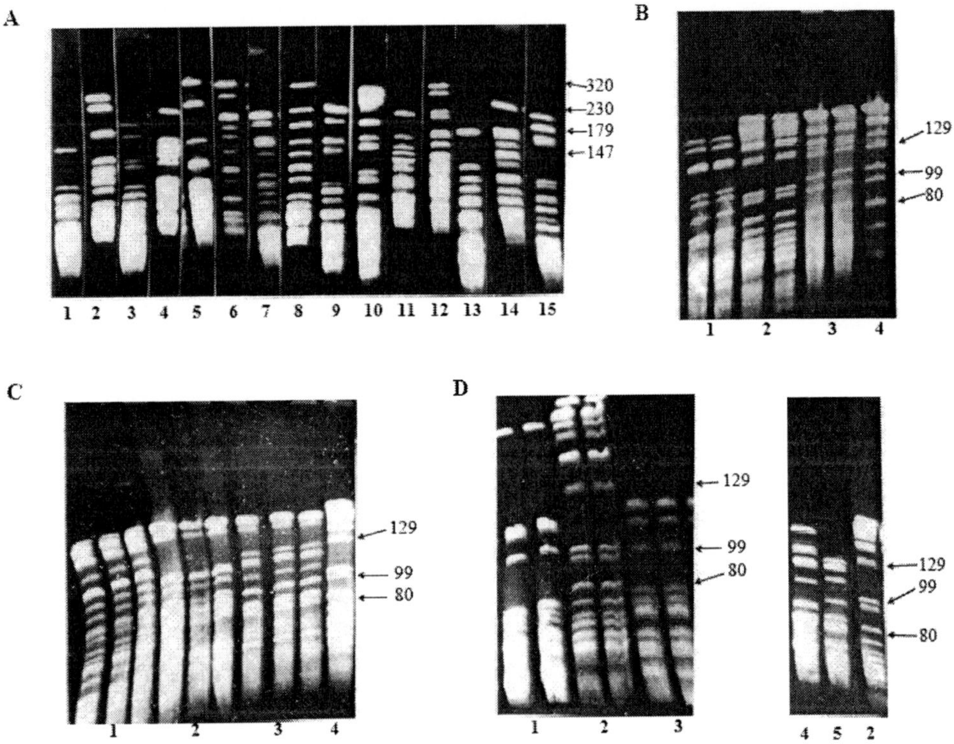

Fig.1. Polymorphism of the chromosomal DNA structure from the strains of acidophilic chemolithotrophic microorganisms: **(A)** *A. ferrooxidans* (1-15); **(B)** *S. sibiricus* (1-3), 4 – DNA size marker (*A. ferrooxidans* VKM B-458); **(C)** *A. thiooxidans* (1-3), 4 – DNA size marker; **(D)** *F. acidiphilum* (1, 3-5), 2 – DNA size marker. Electrophoresis conditions: **(A)** 120 V, 13°C, 10-s pulse, 44-h run; **(B-D)** 130 V, 13°C, 10-s pulse, 68-h run. At the side of lanes, the sizes of the DNA fragments in kb are given.

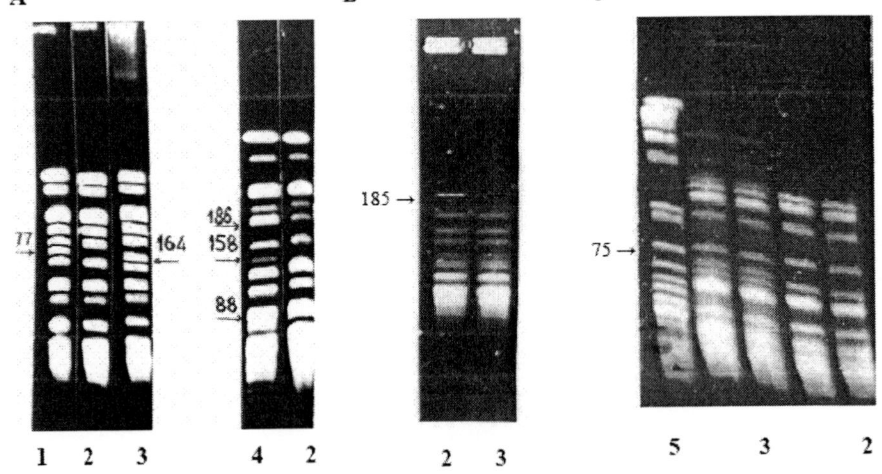

Fig. 2. Changes in the chromosomal DNA structure in the strains of acidophilic chemolithotrophs **(A)** *A. ferrooxidans* TFBk , **(B)** *S. thermosulfidooxidans* sp. *asporogenes* and **(C)** *F. acidiphilum* adapted to growth on media with novel oxidation substrate: 1 – FeS_2, 2 – Fe^{2+}, 3 – S^0, 4 – concentrate, 5 – DNA size marker.

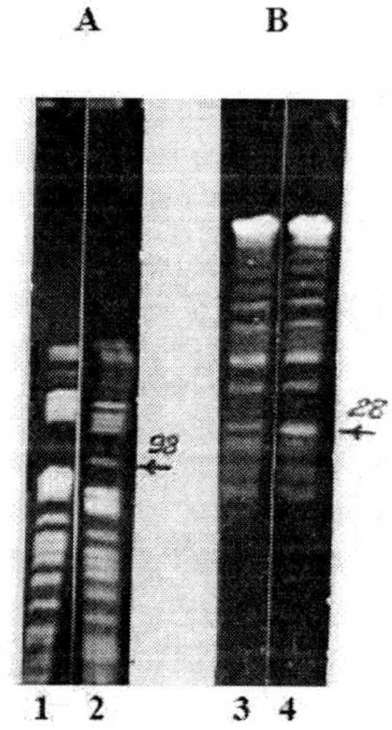

Fig. 3. Changes in the chromosomal DNA structure in *A. ferrooxidans* strains TFZ (2) and TFAs2 (4) adapted to high **(A)** zinc and **(B)** arsenic concentrations in growth media. Initial parent strains: VKM B-458 (1), TFY (3).

Pyrite is the most widely distributed sulfide mineral on the Earth. Stoichiometrically, pyrite (FeS_2) contains 46.6% Fe and 53.4% S. However, the ratio of elements can vary from $FeS_{1.78}$ to $FeS_{2.2}$ or from $Fe_{1.11}S_2$ to $Fe_{0.9}S_2$. Physico-chemical properties of pyrites used in our work and conventionally named "pyrite 5" and "pyrite 6" are presented in Table 1. The pyrites were of different origin and differed in density (mean density of pyrites is 4.9-5.2 g/cm^3), microhardness, impurity concentration, surface morphology, crystallochemical peculiarities and electrophysical properties: the conductivity type, the values of resistivity and TEP. A comparison between two pyrites allow one to suggest that pyrite 6 should be more susceptible to chemical and bacteriochemical oxidation. Pyrite 6 is characterized by greater homogeneity of microhardness, higher level of surface structure imperfections, higher concentrations of sulfur and impurities that impair the crystal lattice.

We compared the course of adaptation of two *A. ferrooxidans* strains to the oxidation of two pyrites in a diluted pulp (S:L = 1:100). After four passages both strains achieved the limit of adaptation. Strain TFBk grew better on both pyrites than strain TFV-1 did. This could be probably explained by the prehistory of strain TFBk isolated from a complex multicomponent substrate, which favoured the development of labile system for the regulation of switching metabolism from the oxidation of one substrate to another. Strain TFBk also oxidized both pyrites more actively than strain TFV-1 did.

Table 1. Physicochemical characteristics of pyrites

Characteristic	Pyrites	
	5	6
Physical properties:		
Density, g/cm^3	5.6	5.72
Microhardness, kgs/mm^2	1360 – 1780 (1458) 794 – 824 (802)	1119 - 1233 (1162)
Impurities	Sb, V, Ni	Co, Ni, As, Ag
Surface peculiarities	Pyrite grains had uneven step-like chips. Some grains had shell-like fractures. Fine inclusions of feldspar were often observed.	Higher density of surface imperfections than in pyrite 5. Cavities and acute-angled points were present. Uniform planes were seldom observed.
Crystallochemical peculiarities of natural samples:		
Concentration, %		
Fe	40.78 – 50.08	40.35 – 44.64
S	49.45 – 59.08	56.27 – 59.09
Co	-	0.21 – 0.62
Ni	0.052 – 0.26	0.204 – 0.37
V	0.032 – 0.16	-
As	-	0.252 – 0.81
Ag	-	0.164 – 0.82
Sb	0.084 – 0.42	-
Electrophysical properties:		
Conductivity type	n Electron type	p Hole type
Resistance value, lnR (R, ohm)	from 4.77 to 9.24	from 7.15 to 9.35
TEP value, mcV/ ^{0}C	from -203.72 to -16.33	from 110.16 to 274.22
Reaction with HNO_3 (0,67N) pH 0.95, 18 ^{0}C	Better soluble than pyrite 6 (contains less sulfur)	Less soluble than pyrite 5 (contains more sulfur)

Unexpected difficulties emerged when chemical and bacteriochemical pyrite oxidation was conducted in dense pulps. Under the industrial processing of ores and concentrates, pyrites, mixed with other sulfide minerals, are oxidized in dense pulps at S:L = 1:5 and at even higher pulp density. In our experiments employing a monomineral fraction, *A. ferrooxidans* TFV-1 did not develop and even died on media with either type of pyrite at S:L = 1:20 (Tables 2 and 3). The substitution of the liquid phase with a fresh medium did not furnish the positive result. At the same time, reinoculation of the liquid phase by strain TFV-1 provided normal growth at the expense of Fe^{2+} oxidation. Thus, the liquid phase on its own, without the solid phase, did not inhibit growth of bacteria. The increase in the Eh value through the addition of Fe^{3+} stimulated growth of strain TFV-1 on pyrite 6, but not on pyrite 5.

Table 2. Bacteriochemical oxidation of pyrite 5 at S: L = 1: 20 by strain *A. ferrooxidans* TFV-1

Days	Growth and oxidation parameters				
	pH	**Eh, mV**	**Fe^{3+}, g/l**	**Fe^{2+}, g/l**	**Cells per ml**
0	1.88	590	0,28	0.42	$1.1x10^7$
2	1.79	550	0,14	0.77	0
Repeated inoculation up to $1x10^7$ cells/ml					
10	1.46	570	0.40	2.23	0
pH value adjusted to 1.96					
16	1.60	570	0.35	3.91	0
Substitution of liquid phase with fresh medium, inoculation up to $1x10^7$ cells/ml					
36	1.42	550	0.63	2.80	0
Addition Fe^{3+} (5.93 g/l), increase of Eh up to 650 mV, pH value adjusted to 1.88					
43	1.34	600	2.09	6.28	0
Repeated inoculation up to $1x10^7$ cells/ml, pH value adjusted to 1.86					
45	1.72	590	1.26	5.58	0
Addition Fe^{3+} (5.72 g/l), increase of Eh up to 620 mV, pH value adjusted to 1.94					
47	1.4	570	2.79	5.03	0

Table 3. Bacteriochemical oxidation of pyrite 6 at S: L = 1: 20 by strain *A. ferrooxidans* TFV-1

Days	Growth and oxidation parameters				
	pH	**Eh, mV**	**Fe^{3+}, g/l**	**Fe^{2+}, g/l**	**Cells per ml**
0	1.94	600	0.28	0.35	$4.39x10^7$
2	1.77	540	0.14	0.91	0
Repeated inoculation up to $1x10^7$ cells/ml					
10	1.42	560	0.42	3.49	0
pH value adjusted to 2,0					
16	1.63	560	0.48	5.3	0
Substitution of liquid phase with fresh medium, inoculation up to $1x10^7$ cells/ml					
36	1.33	580	0.84	3.6	0
Addition Fe^{3+} (4.88 g/l), increase of Eh up to 643 mV, pH value adjusted to 1.67					
43	1.33	670	7.4	2.4	$3.59x10^7$
pH value adjusted to 1.84					
47	1.4	640	4.6	0.56	$2.05x10^8$
pH value adjusted to 2.47					
54	1.45	710	2.44	0.07	$3.56x10^8$

During the chemical oxidation (without bacteria) of both pyrite types, Fe^{2+} was accumulated, which indicated the electrochemical oxidation of pyrites (Table 4). It is remarkable that in the presence of bacteria, until the addition of Fe^{3+}, much more Fe^{2+} was accumulated during the oxidation of pyrite 6, than in the absence of bacteria (Tables 3 and 4).

Table 4. Chemical oxidation of two types pyrite

Days	Pyrites							
	5				6			
	Oxidation parameters							
	pH	Eh, mV	Fe^{3+}, g/l	Fe^{2+}, g/l	pH	Eh, mV	Fe^{3+}, g/l	Fe^{2+}, g/l
0	1.7	560	0.21	0.56	1.9	510	0.07	0.78
5	1.8	540	0.07	1.26	1.9	530	0.07	0.84
10	1.7	550	0.07	1.33	1.8	540	0.07	1.26
21	1.6	560	0.14	2.23	1.7	550	0.14	2.00
40	1.5	560	0.14	3.00	1.6	550	0.14	2.51

The results obtained allowed us to suggest that the main factors preventing the development of strain TFV-1 in the medium with pyrite 6 at the pulp density S:L = 1:20 were the electrochemical interactions between the liquid phase (Eh) and the solid phase – electrode potential of pyrite (EP). It is evident that the similarity of the values of the medium Eh and EP of pyrite 6 eliminated the possibility of using pyrite 6 as a source of energy and, hence, of the bacterial growth. Similar results were reported earlier [7]. However, the absence of growth of strain TFV-1 in like experiments in the medium with pyrite 5 is yet to be explained. It may be suggested that the Eh value of the medium after the addition of Fe^{3+} was still not high enough to initiate the bacteriochemical oxidation of pyrite 5.

Considering the above data, further experiments with another strain TFBk were conducted at S:L = 1:50, i.e., in a more diluted pulp. Under such conditions, strain TFBk started to grow and oxidize pyrites of both types only after the addition of Fe^{3+}, which increased the Eh value of the medium and, hence, the difference between the Eh and EP of the pyrites (Tables 5 and 6). Ores and concentrates always contain various sulfide minerals that form electrochemical pairs, where one mineral plays a role of anode and is oxidized, while another one acts as a cathode. For example, arsenopyrite is the first to be oxidized in a pair pyrite – arsenopyrite. In a monomineral fraction of the pulp at S:L > 1:50, the equilibrium conditions are set up, preventing the pyrite oxidation.

One of the valuable theoretical approaches in the investigation of the physico-chemical conditions favourable for the oxidation of sulfide minerals lies in studying their electrophysical properties: conductivity type, the values of resistance and TEP. Analysis of pyrites after their chemical and bacteriochemical oxidation revealed the changes in their electrophysical properties as compared with the initial characteristics. The most essential changes were found after chemical oxidation: in pyrite5, with initial n-type of conductivity and negative values of TEP, the inversion to p-type of conductivity with positive values of TEP was observed, i.e., the electron conductivity changed to hole conductivity; in pyrite 6, a significant decrease of the TEP value both at S:L = 1:20 and at S:L = 1:50 was observed. The inversion of conductivity type may be connected with the increase in the concentration of acceptor impurities in anionic group. The logarithm of the resistance value, lnR, decreased to a greater extent at S:L = 1:20 than at S:L = 1:50, i.e., the conductivity has increased. Thus, the structure of pyrites – half-conductors after their chemical oxidation in the medium with low pH value, became more of a conductor type. Conceivably, this is related to the leaching of impurity elements.

Table 5. Bacteriochemical oxidation of pyrite 5 at S: L = 1: 50 by strain *A. ferrooxidans* TFBk

Days	Growth and oxidation parameters				
	pH	Eh, mV	Fe^{3+}, g/l	Fe^{2+}, g/l	Cells per ml
0	1.96	560	0.21	0.14	$1.3x10^7$
10	1.84	570	0.28	1.64	0
Repeated inoculation up to $1x10^7$ cells/ml					
13	1.65	570	0.56	2.09	0
Addition Fe^{3+} (2.34 g/l), increase of Eh up to 600 mV					
20	1.43	660	5.58	1.35	$2.66x10^7$
pH value adjusted to1.95					
22	1.39	610	3.91	0.98	$2.2x10^8$
pH value adjusted to1,98					
23	1.50	690	4.05	0.07	$2.83x10^8$

Table 6. Bacteriochemical oxidation of pyrite 6 at S: L = 1: 50 by strain *A. ferrooxidans* TFBk

Days	Growth and oxidation parameters				
	pH	Eh, mV	Fe^{3+}, g/l	Fe^{2+}, g/l	Cells per ml
0	1.98	570	0.28	0.21	$1.37x10^7$
7	1.93	550	0.14	0.98	0
Repeated inoculation up to $1x10^7$ cells/ml					
10	1.86	560	0.21	1.19	0
Addition Fe^{3+} (2.16 g/l), increase of Eh up 640 mV					
17	1.42	720	6.7	0	$3.05x10^8$
pH value adjusted to1.88					
19	1.44	660	7.54	0.07	$8.29x10^8$
pH value adjusted to 1.95					
21	1.43	770	5.86	0	$1.08x10^9$

Electrophysical characteristics of the pyrite samples were less changed after the bacteriochemical treatment. In pyrite 5 treated in the medium with S:L = 1:20 inoculated by strain TFV-1, the TEP value slightly decreased while the conductivity remained almost the same. Under the same conditions, the TEP value of pyrite 6 decreased, the conductivity was scarcely affected. After the treatment in the medium with S:L = 1:50 inoculated by strain TFBk, a slight decrease the TEP value and conductivity was observed in both types of pyrites.

It may be suggested that such difference in the effect of chemical and bacteriochemical oxidation on the electrophysical characteristics of pyrites is due to the fact that bacteria more efficiently oxidize thermodynamically weak zones, therewith leveling the changes occurring during chemical oxidation.

We studied the influence of physical, crystallochemical and electrophysical peculiarities of pyrites on the genomic characteristics and variability of strains under consideration. After multiple passages (>10) of strain TFV-1 on medium containing pyrite 5 at S:L = 1:100, we revealed the changes in the chromosomal DNA structure (Fig. 4). When pyrite 6 was used as the oxidation substrates, no changes in the DNA structure were observed. It is interesting that it was precisely pyrite 5 that was not oxidized by strain TFV-1 at S:L = 1:20 and was less efficiently oxidized at S:L = 1:100, than pyrite 6 was. It may be suggested that certain changes in the genome of strain TFV-1 are required for the realization of its oxidative capacity. The adaptation of strain TFV-1 to growth on medium containing pyrite 5 or 6 caused changes in their plasmid profiles (fig. 5). Earlier, changes in the plasmid profiles of strain TFV-1 were noted after the adaptation of this strain to other oxidation substrates [8]. The number and size of plasmids varied depending on the type of pyrite employed. The adaptation of strain TFBk to growth on either of pyrites caused changes in the regions corresponding to small fragments of the chromosomal DNA (Fig. 4). No changes in the plasmid profiles of this strain were found (Fig.5). Thus, it was for the first time demonstrated that the changes in chromosomal DNA structure may result not only from switching of the metabolism to the oxidation of a new energy substrate but may also depend on the physical, crystallochemical and electrophysical peculiarities of one and the same substrate.

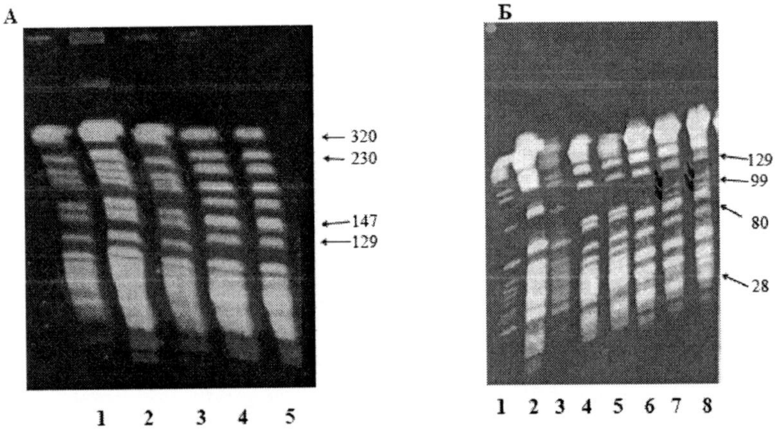

Fig. 4. Restriction patterns of the chromosomal DNA from *A. ferrooxidans* strains. **(A)** Electrophoresis conditions: 120 V, 13°C, 10-s pulse, 44-h run. Lanes: 1 – strain TFV-1 grown on medium with Fe^{2+}; 2 – strain TFV-1 grown on medium with pyrite 5; 3 – strain TFV-1 grown on medium with pyrite 6; 4 and 5 – strain VKM B-458 grown on medium with Fe^{2+}. **(B)** Electrophoresis conditions: 130 V, 13°C, 10-s pulse, 68-h run. Lanes: 1 – strain TFV-1 grown on medium with Fe^{2+}; 2 – strain TFV-1 grown on medium with pyrite 5; 3 – strain TFV-1 grown on medium with pyrite 6; 4 and 5 – strain VKM B-458 grown on medium with Fe^{2+}; 6 – strain TFBk grown on medium with Fe^{2+}; 7 – strain TFBk grown on medium with pyrite 5; 8 – strain TFBk grown on medium with pyrite 6. At the side of lanes, the sizes of the DNA fragments in kb are given.

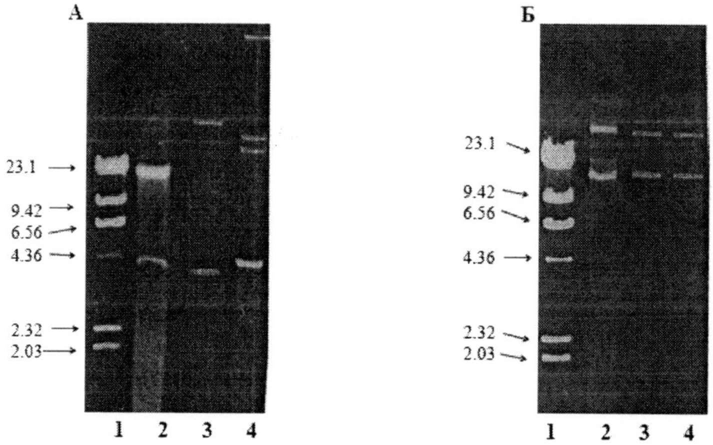

Fig. 5. Plasmid profiles of *A. ferrooxidans* strains TFV-1 and TFBk adapted to pyrites 5 and 6. **(A)** 1 – phage lambda DNA cleaved by restriction endonuclease *Hind*III; 2 – strain TFV-1 grown on medium with Fe^{2+}; 3 – strain TFV-1 grown on medium with pyrite 5; 4 – strain TFV-1 grown on medium with pyrite 6. **(B)** 1 – phage lambda DNA cleaved by restriction endonuclease *Hind*III; 2 – strain TFBk grown on medium with Fe^{2+}; 3 – strain TFBk grown on medium with pyrite 5; 4 – strain TFBk grown on medium with pyrite 6. At the side of lanes, the sizes of the DNA fragments in kb are given.

Growth temperature and pH of the medium are also important factors that may cause changes in acidophilic chemolithotrophs. We studied the effect of lowering the pH value of the medium in different temperature regimes (4-6°C, 20-22°C and 28-30°C) on the activities of growth and Fe^{2+} oxidation and also on the chromosomal DNA structure in three strains of *A. ferrooxidans,* TFBk,TFUd and TFMt, isolated from different substrates and climatic zones. Strain TFBk was isolated from gold-bearing pyrite-arsenopyrite ore concentrate from the Kazakhstan deposit, strain TFUd – from copper ore obtained in permafrost zone of Chitinskaya oblast in the southern Siberia and strain TFMt – from gold-arsenic pyrite-arsenopyrite ore obtained in permafrost zone of Magadanskaya oblast.

The possibility to organize bacteriochemical processes for dump leaching of nonferrous metals in permafrost zones is one of the pressing problems of today. It is, therefore, very important to know the adaptation limits of *A. ferrooxidans* strains to low pH values within different temperature ranges. It is known that the lower is the pH value of the leaching solution, the lower may be the temperature for running the leaching process. Table 7 shows the kinetic parameters of growth and Fe^{2+} oxidation by three strains in three temperature regimes and under experimental lowering of pH values of the medium. Regardless of the strain employed, the site it was isolated from, or the temperature regime of the experiment, the lowest possible pH value allowing the growth and Fe^{2+} oxidation was 0.9. Strain TFBk showed a higher growth rate than other strains in the temperature range from 4 to 20°C, strain TFUd – from 28 to 30°C. At low temperatures, kinetic growth parameters μ_{max} and t_d of strain TFMt were close to those of strain TFUd. However, in the temperature range 28-30°C, strain TFMt, isolated as well as strain TFUd from pe`rmafrost zone, differed from the latter by the lowest growth activity at low pH values. Thus, no correlation has been found between the phenotypic peculiarities of *A. ferrooxidans* strains and the site of their isolation.

Table 7. Growth kinetic parameters of *A. ferrooxidans* strains at lowered pH values in different temperature regimes

Strains	Temperature regimest, ^0C	pH								
		1.1			1.0			0.9		
		μ_{max}, h^{-1}	t_d, h	Activity of oxidation Fe^{2+}, g/l·days	μ_{max}, h^{-1}	t_d, h	Activity of oxidation Fe^{2+}, g/l·days	μ_{max}, h^{-1}	t_d, h	Activity of oxidation Fe^{2+}, g/l·days
TFBk	4 - 6	0.013	53.3	0.7	0.013	53.3	0.5	0.013	53.3	0.4
	20 - 22	0.020	34.7	1.2	0.020	34.7	0.9	0.014	49.5	0.6
	28 - 30	0.026	26.7	1.1	0.012	57.8	0.5	0.009	77.0	0.2
TFUd	4 - 6	0.009	77.0	0.5	0.008	86.6	0.5	0.011	63.6	0.3
	20 - 22	0.018	38.5	1.2	0.017	40.8	0.9	0.014	49.5	0.4
	28 - 30	0.034	20.4	1.3	0.024	28.9	1.3	0.019	36.5	0.7
TFMt	4 - 6	0.010	69.3	0.7	0.010	69.3	0.4	0.010	69.3	0.2
	20 - 22	ND	ND	ND	ND	ND	ND	ND	ND	ND
	28 - 30	0.006	115.5	ND	0.005	138.0	ND	0.005	138.6	ND

Note: ND stands for «not determined»

We found no changes in the chromosomal DNA structure of strains either adapted to growth at low pH values, or cultivated in different temperature ranges. Evidently, these two factors can play an evolutionary or taxonomic role not at the strain level, but at least at the level of a species. In the species *A. ferrooxidans*, the standard response to changes of the medium pH and the temperature was within the limits employed in our experiments.

Thus, of all medium parameters considered in this work: oxidation substrate, concentration of heavy metal ions and toxic elements, suboptimal values of pH and temperature, it was the switching of metabolism to the oxidation of a new energy source that caused the highest genotypic variability of acidophilic chemolithotrophs; as stated above, the genotypic characteristics were affected not only by the chemical composition of a new oxidation substrate but also by its physical, crystallochemical and electrophysical properties.

ACKNOWLEDGEMENTS

This work was supported by grants from the Russian Foundation for Basic Research (RFBR) "Leading Scientific Schools", project no. 1742.2003.4, the RFBR Initiative Research Project, project no. 01-04-48270, 03-04-49294 and Goskontrakt, project no. 43.073.1.1.25.15.

REFERENCES

1. T.F. Kondratyeva, L.N. Muntyan, G.I. Karavaiko: *Microbiology (UK),* **141** (5), 1157 (1995).
2. T.F.Kondrat'eva, V.S. Melamud, I.A. Tsaplina, T.I. Bogdanova, A.A. Senyushkin, T.A. Pivovarova, G.I. Karavaiko: *Microbiology,* **67** (1), 19 (1998) (in Russian).
3 O.V. Golyshina, T.A. Pivovarova, G.I. Karavaiko, T.F. Kondratyeva, E.R.B. Moor, W.-R. Abraham, H. Lunsdorf, R.N. Timmis, M.M. Yakimov, P.N. Golyshin: *Int. J. Syst. Evol. Microbiol.,* **50** (3), 997 (2000).
4 T.F. Kondrat'eva, S.N. Ageeva, L.N. Muntyan, T.A. Pivovarova, G.I. Karavaiko: *Microbiology,* **71** (3), 373 (2002).(in Russian).
5 S.N. Grudev: *Biogeotechnology of metals.* Centre for international projects GKNT, Moscow, 1985. 435 p. (in Russian).
6 G.I. Karavaiko, L.S. Smolskaya, O.K. Golyshina, M.A. Yagovkina, E.Y. Egorova: *Fuel Processing Technology,* **40** ( ), 151 (1994).(in Russian).
7 G.I. Karavaiko: *Biogeotechnology of metals.* Manual. Centre for international projects GKNT, Moscow, 1989. 350 p. (in Russian).
8 S.N. Ageeva, T.F. Kondrat'eva, G.I. Karavaiko: *Microbiology,* **72** (5), (2003). (in Russian).

Chapter 18

CHEMOLITHOTROPHIC BACTERIAL COMMUNITY IN THE PULP OF A PILOT PLANT DURING GOLD-ARSENIC PYRRHOTITE CONCENTRATE PROCESSING

T.F. Kondrat'eva [1]**, *V.S. Melamud*** [1]**, *R.Ya. Aslanukov*** [2]**,**
O.B. Voronina [2] ***and G.I. Karavaiko*** [1*]

[1] Institute of Microbiology, Russian Academy of Sciences, Moscow, 117312 Russia
[2] Polus Gold Mining Company, Krasnoyarsk, 660061 Russia

ABSTRACT

This study has a focus on the community of chemolithotrophic bacteria and distribution of its particular representatives in a dense pulp during processing of gold-arsenic pyrrhotite concentrate under semi-industrial conditions at 30^0 C in the Olimpiadinskaya gold-mining factory. It was shown that a continuous pilot plant, consisting of a cascade of reactors, functions as a stable straight-flow system with permanently changing conditions in reactors, providing for the adaptation and high activity of bacterial communities. Under processing at 30^0C, two major native aboriginal chemolithotrophic communities were found, including (i) mesophilic strains of *Acidithiobacillus ferrooxidans* TFO and *Acidithiobacillus thiooxidans* TTO with the growth optimum at 30^0 C and (ii) moderately thermophilic strains of *Sulfobacillus* sp. (Kr 1), *Leptospirillum ferrooxidans* LFO, and *Ferroplasma acidiphilum* Y-2 with the growth optima at 40, 45, and 50^0C, respectively. The most part of bacteria (93.5 – 97.5%) was found in the liquid phase of the processed pulp. The number of cells firmly attached to concentrate particles account for $3.2x10^9$ and $3.17x10^9$ in reactors 1 and 4, respectively, or 7.6% and 2.3% of the total cells in the solid phase. The role of attached and freely floating bacterial cells in the establishment of physicochemical conditions and oxidation of sulfide minerals and sulfur in the pulp is discussed.

Keywords: community of chemolithotrophic bacteria, Acidithiobacillus ferrooxidans, Acidithiobacillus thiooxidans, Sulfobacillus, Leptospirillum ferrooxidans, Ferroplasma, gold-arsenic concentrate

---

[*] Author for correspondence. E-mail: gregor@inmi.host.ru

INTRODUCTION

To date, gold-mining industry has increasingly exploited recalcitrant ores and concentrates produced on their basis, in which noble metals are mainly included in the crystal lattice of sulfide minerals, such as arsenopyrite (FeAsS) and pyrite (FeS_2). Commonly applied technologies are either unsuitable for processing of recalcitrant ores and concentrates (because of low efficiency of gold and silver recovery) or ecologically hazardous.

The use of the biohydrometallurgical technology that relies on the oxidation of sulfide minerals in reactors by microorganisms allows noble minerals to be liberated and then extracted from the cake by hydrometallurgical methods. This technology has already found practical application in commercial plants in South Africa, Brazil, Ghana, Australia, USA, and Russia. The scale of processing gold-arsenic concentrates in different plants varies from 40 to 1200 tons per day [1].

Abroad, the biohydrometallurgical technology commonly uses the community of thermotolerant Gram-negative bacteria *Acidithiobacillus caldus* and *Leptospirillum ferrooxidans* developing growing at 45^0C [2].

This work was aimed at studying the community of chemolithotrophic bacteria and regularities of the distribution of its members in a dense pulp during gold-arsenic pyrrhotite concentrate processing at 30^0C under semi-industrial conditions at the Olimpiadinskaya gold-mining factory.

MATERIALS AND METHODS

The main sulfide minerals in the concentrates utilized at Olimpiadinskaya gold-mining factory are pyrrhotite (~ 25.3%), pyrite (~ 6.4 %), arsenopyrite (7.2 %), and antimonite (1.2 – 1.5 %); other sulfide minerals occurred in minor amounts. Bioleaching experiments were conducted in a pilot plant consisting of a a contact reactor and five consecutively connected working reactors, each 2.8 m^3 in volume (Fig. 1). The contact reactor served to prepare a pulp with a prespecified density and to regulate its mineral composition and pH. Tests were carried out over three years under continuous conditions at a constant pulp density of ~ 20% (solid : liquid phase = 1:5) with pneumomechanical agitation at a rate of ~300 rpm. Air was fed to the facility from a compressor station in a proportion of 1 m^3 per 1 m^3 of the pulp.

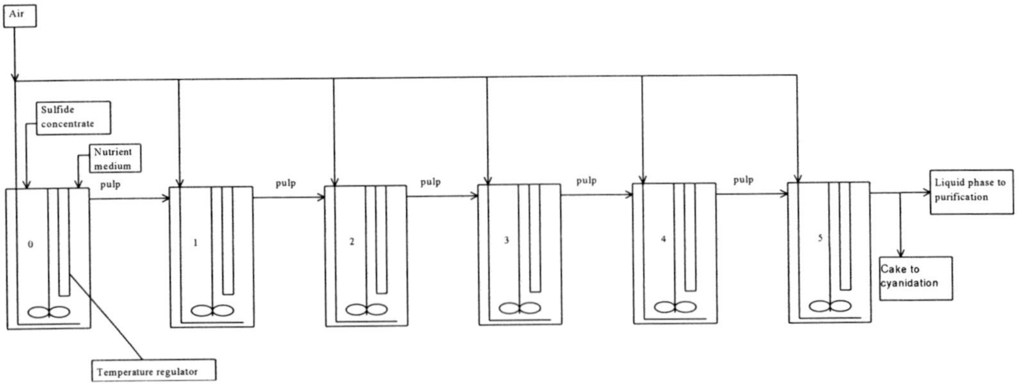

Fig. 1. Basic scheme of the pilot plant used for tank bacterial oxidation of gold-containing sulfide concentrate. The plant consists of a contact reactor (0) and five serially connected reactors (1-5).

Analysis of the Bacterial Community in the Pulp

The total number of bacteria in the community was counted under a phase-contrast microscope. To analyze the bacterial species composition and to determine the number of viable cells, the method of serial decimal dilutions in specific nutrient media [3] was used with subsequent isolation of pure cultures. The isolated bacterial strains were identified by their phenotypic features, DNA-DNA homology [4], and 16S rRNA gene sequencing [5]. The strain polymorphism of the studied community was studied by pulsed-field gel electrophoresis of chromosomal DNA digested by restriction endonucleases as described earlier [6]. The potential activity of the bacterial community was assessed manometrically using the Warbourg instrument after adding 5 ml of the pulp and 3 g/l of Fe^{2+}; the incubation was for 15 min at 30^0C.

Analytical Methods

The contents of Fe^{2+} and Fe^{3+} ions were determined by titration with Trilon [7]. The values of pH and Eh were measured using an I-130.2 ion meter (model M.1); the Eh values were corrected with account for a reference H^+ electrode. The content of elements in the solid and liquid phases of the pulp was analyzed in a laboratory of the Olimpiadinskaya gold-mining factory using the conventionally accepted methods.

Results and Discussion

Microbiological and Chemical Analysis of the Pulp

In the second year of testing, the reactors were examined three times over a month. Average results of total number determinations, respiratory activity assessments, pH and Eh measurements, and Fe^{2+} and Fe^{3+} contents are given in Fig.2. The total bacterial cell number was found to be surprisingly constant in the pulp of reactors over all the period of observations and varied in a range $5.0x10^9 - 8.2x10^9$ ml^{-1}. The pH value gradually declined in the last reactors from 2.0 to 1.3-1.4 or even to 1.15 when elemental sulfur was more actively oxidized (Table 1).

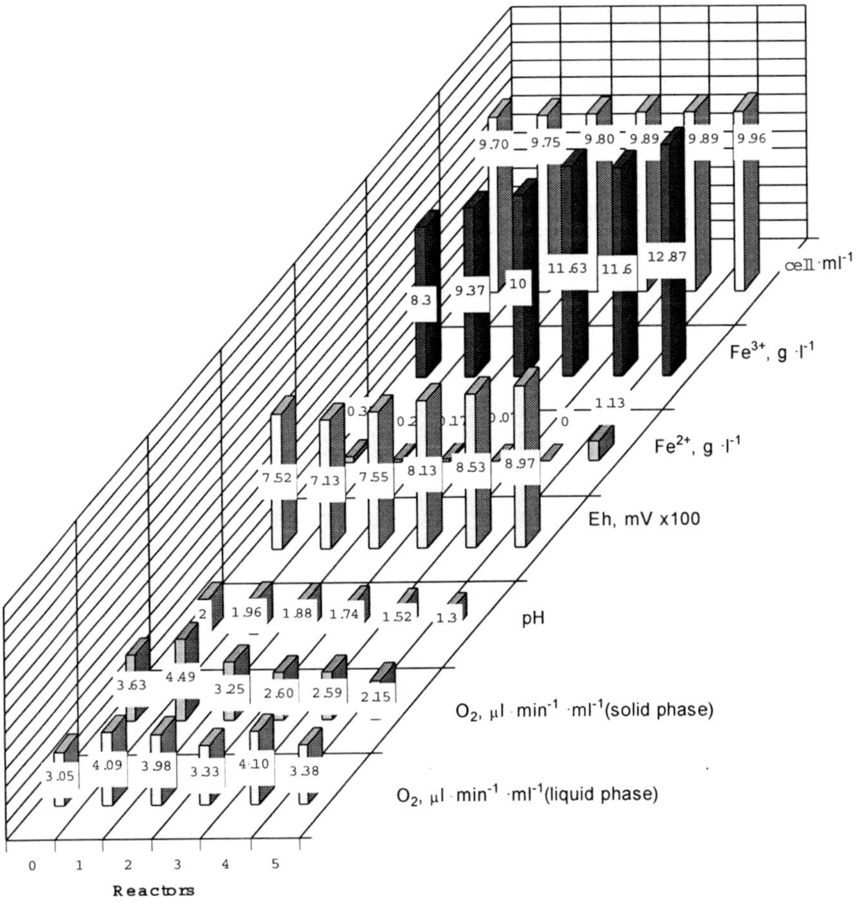

Fig. 2. Microbiological and physicochemical characteristics of the pulp in the reactors (SP, solid phase; LP, liquid phase).

Table 1. Contents (%) of sulfur species in the solid phase of pulp

Components	Reactors			
	0	1	3	4
Total sulfur	16.06	14.49	14.01	10.41
Sulfide sulfur	11.17	10.3	8.23	5.91
Elemental sulfur	2.38	2.45	1.90	1.84

The redox potential (Eh) of the pulp was high in all the reactors and increased beginning from the first reactor to a value as high as 900 mV, testifying to strongly oxidative conditions.

In contrast to the low Fe^{2+} content in the pulp (0-1.13 g/l), the concentration of Fe^{3+} ions was high and increased from 8.3 to 12.3 g/l from the zero (contact) to the fifth reactor. The As^{3+} content in the liquid phase of the pulp was relatively small and varied in a range of 0.8 – 2.5 g/l, rising in the last reactors, whereas the concentration of As^{5+} ions did not exceed 0.007

– 0.014 g/l. A major part of pentavalent arsenic was precipitated with the formation of ferric arsenates ($FeAsO_4$).

As judged from the respiratory activity of chemolithotrophic bacteria measured by the O_2 uptake during Fe^{2+} oxidation, there were some differences both between the reactors and between the liquid and solid pulp phases. For instance, the respiratory activity of bacterial cells was remarkable in reactors 1 and 2 and declined in the following reactors. In the first and second reactors, the O_2 uptake was similar in the liquid and solid phases. However, beginning with reactor 3, bacteria were more active in the liquid phase of the pulp and likely specialized oxidizing elemental sulfur and its water-soluble compounds. Chemical analyses of the solid phase of the pulp (Table 1) also pointed to a decrease in the contents of sulfur and sulfur-containing compounds with a low degree of oxidation in the third and further reactors. The degree of arsenopyrite and pyrite oxidation reached 96-99% and 78-80% respectively.

Hence, the tests proved that the used continuous pilot facility functioned as a stable system with sequentially changing conditions in the reactors, providing for adaptivity and high activity of bacteria and thereby profound oxidation of sulfide minerals in the pulp and liberation of noble metals.

Biodiversity of Chemolithotrophic Bacteria in the Pulp Community

In the microbial community developing in the pulp, we found and identified *A. ferrooxidans* (strain TFO) and *A.thiooxidans* (strain TTO) with growth optima at 30^0C, as well as *L.ferrooxidans* (strain LFO) and *Sulfobacillus sp.* (strain Kr 1) with growth optima at 45 and 40^0C, respectively. In addition, we revealed mesophilic archaea of the genus *Ferroplasma,* oxidizing Fe^{2+} and FeS_2 at 30-35°C and close to previously described microorganisms [8, 9], as well as yet unidentified acidophilic organotrophic bacteria. The data on their number in the microbial community studied are given in Fig.3.

Two microbial communities of the pulp stood out for their cell number, namely *A.ferrooxidans* - *A.thiooxidans* and *Sulfobacillus* sp. - *L.ferrooxidans*. At 30^0C, the first community was dominant, and this temperature was evidently optimal for its development. The number of *A.ferrooxidans* and *A.thiooxidans* cells varied in the contact and fifth reactor in the ranges of 3.0×10^8 to 2.5×10^9 ml^{-1} and 9.5×10^8 to 2.5×10^9 ml^{-1}, respectively (Fig. 3).

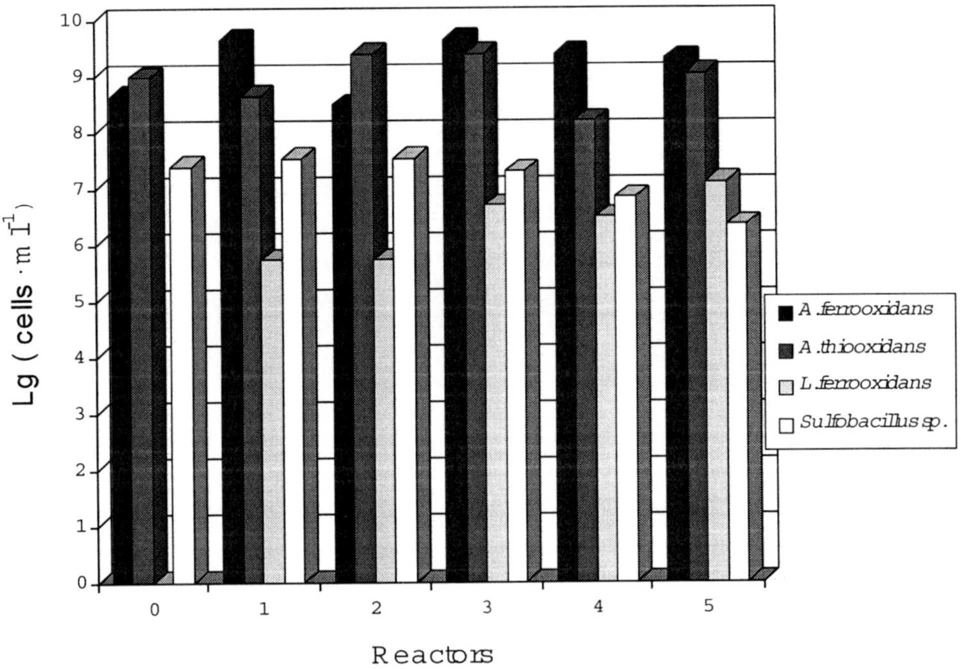

Fig. 3. Cell numbers of chemolithotrophic bacteria in different reactors.

The ability of the two communities to adapt to various substrates in the pulp of the reactors was estimated from their growth activities after the inoculation of serially diluted pulp samples into 9K medium containing Fe^{2+} ions and Waksman medium containing elemental sulfur [3]. After inoculation of the Fe^{2+} - containing medium with pulp samples, the maximum number of *A. ferrooxidans* cells was revealed on the 11th -17th days of cultivation, irrespectively of the reactor from which the samples were taken. This points to a great activity of *A. ferrooxidans* in the oxidation of Fe^{2+} and sulfide minerals. The maximum number of *A. thiooxidans* cells present in the pulp of any of the reactors could be revealed on medium with S^0 only after 27 days of cultivation. On the 11th day, bacterial growth on medium with elemental sulfur corresponded to the cell number of $4.5x10^6$ if the inoculum was taken from reactors 3 - 5 and only to $2.5x10^2$ - $2.5x10^4$ cells ml^{-1} if the inoculum was taken from reactors 0 - 2, although the actual content of the corresponding bacteria in the pulp was 1.15 - $2.5x10^9$ cells ml^{-1} (in reactors 3 - 5) and $9.5x10^8$; $4.5x10^8$ and $2.5x10^9$ ml^{-1} (in reactors 0 - 2) (Table 2). These data provide evidence for the functional adaptive variability of *A. thiooxidans* in reactors with changing conditions. A more active sulfur oxidation occurred beginning with from reactor 3; this conclusion is supported by the data of chemical analyses of pulp samples (Table 1).

Table 2. The growth dynamics of *A. thiooxidans* in the medium with elemental sulfur after inoculation with the pulp samples from reactors.

No of reactor	Cell number (ml^{-1}) after inoculation for		
	11 days	17 days	27 days
0	$2.5x10^4$	$4.5x10^6$	$9.5x10^8$
1	$2.5x10^2$	$9.5x10^6$	$4.5x10^8$
2	$2.5x10^2$	$9.5x10^6$	$2.5x10^9$
3	$1.5x10^7$	$9.5x10^7$	$2.5x10^9$
4	$1.5x10^7$	$1.5x10^7$	$1.6x10^9$
5	$4.5x10^6$	$9.5x10^6$	$1.15x10^9$

The moderately thermophilic Gram-positive bacterium *Sulfobacillus* sp. strain Kr 1 was found in the pulp in an amount of $2.2x10^6$ - $3.5x10^7$ ml^{-1}. It was able to oxidize Fe^{2+}, S^{2-}/S^0, and sulfide minerals while growing in the temperature range of 20 - 55^{0}C with an optimum at 45-50^{0}C and pH ~ 2.0. The number of Gram-negative bacteria *L. ferrooxidans* strain LFO, oxidizing Fe^{2+} and FeS_2 at 45^{0}C amounted to $5.0x10^5$ ml^{-1} in reactors 0-2 and to $5.0x10^6$ - $1.3x10^7$ ml^{-1} in the other reactors. The cell number of *Ferroplasma* sp. oxidizing Fe^{2+} and FeS_2 at 30°C was 10^5 ml^{-1}.

It is difficult to estimate the contribution of each member of the community, as well as of particular communities as a whole, to the oxidative processes occurring in the pulp of the facility under mesophilic conditions. Most probably, *A. ferrooxidans* and *A. thiooxidans*, constituting the first community, play the major role in the oxidation of sulfide minerals and sulfur-containing compounds under the given conditions. However, the second community should not be underestimated, since *L. ferrooxidans* is known to be more resistant at low pH [2] and high concentrations of sulfates [10]. The community of *A. caldus* and *L. ferrooxidans* plays a dominant role in industrial processing of complex gold-arsenic concentrates at 45^{0}C in commercial bioleaching plants in several countries [2]. However, the chemolithotrophic bacterium *A. caldus* was not found in pulps processed at the Olimpiadinskaya gold-mining plant. The community of *L. ferrooxidans* and *A. thiooxidans* actively oxidized various sulfide minerals and dominated during copper leaching at high acidity of the medium (pH 0.7) [11]. *Sulfobacilli* were more active than *A. ferrooxidans* in pyrite oxidation cells at optimal (~ 50^{0}C) [12] and even sub-optimal (30^{0}C) temperatures [13]. Mixed cultures of Gram-positive bacteria or of Gram-positive and Gram-negative bacteria are often more efficient in metal leaching than pure cultures [14, 15]. As to representatives of the genus *Ferroplasma*, their role in gold-arsenic concentrate processing remains as yet unclear.

Upon heating of the pulp due to exothermal oxidative processes, the second community considered (*L.ferrooxidans* - *Sulfobacillus* sp.) may become dominant and determine the operation rate of the reactors. Evidently, the community of mesophilic and moderately thermophilic bacteria is capable of maintaining a high oxidative potential in the wide ranges of temperature and pH in the system of reactors. The role of organotrophic bacteria in biohydrometallurgical processes may consist in the oxidation of organic exometabolites produced by chemolithotrophic bacteria.

Bacterial Genome Studies

In biohydrometallurgical practices, microbiological analysis is usually confined to the species level. However, each type of ore has its own aboriginal strain of chemolithotrophic bacteria [16]; therefore, it is important to detect and monitor these strains in reactors with different physicochemical conditions.

The analysis of the chromosomal DNA structure in *A. ferrooxidans* cells isolated from reactors 1 and 3 pointed to the presence of the same dominant strain (Fig. 4) that was absent from the initial inoculum used at the initiation of the pilot bioleaching facility.

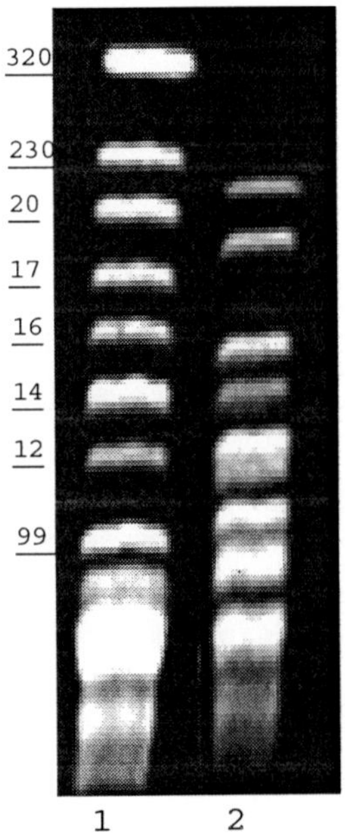

Fig. 4. *XbaI*-restriction pattern of chromosomal DNA from *A. ferrooxidans* strain TFO (2), size marker (1). Conditions of electrophoresis: voltage, 120 V; pulse time, 25 s; duration, 44 h. Hereafter, fragment sizes are indicated in kb.

The restriction pattern of chromosomal DNA from cells of this strain was identical to that of the wild *A. ferrooxidans* TFO strain isolated from the concentrate studied. This fact bears witness to the predominance in the bioleaching process of an aboriginal *A. ferrooxidans* TFO strain adapted to the ore and concentrate, as well as to the conditions produced upon their oxidation. Likewise, the chromosomal DNA restriction profile of the *A. thiooxidans* strain isolated from reactors 1 and 3 was similar to that of the wild strain TTO (Fig. 5A). However, amplification of a 170-kbp fragment was found in the chromosomal DNA of *A.thiooxidans* cells isolated from reactor 3 (Fig.5B). The presence of such a fragment may be due to the adaptation of *A. thiooxidans* to the complex of sulfur-containing compounds present in the pulp of the third and subsequent reactors, as also evidenced by the higher growth rate and sulfur oxidation activity (Table 2). The new spore-forming bacterium *Sulfobacillus* sp. (Kr 1) differed in the chromosomal DNA structure from the known strains (Fig. 6).

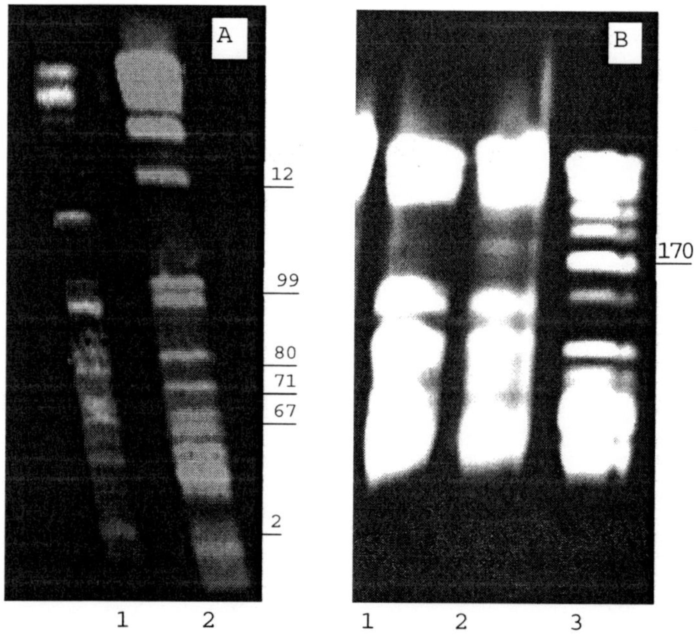

Fig. 5. A. *Xba*I-restriction pattern of chromosomal DNA from *A. thiooxidans* strain TTO(1), size marker (2). Conditions of electrophoresis: voltage, 130 V; pulse time, 10 s; duration, 72 h. B. *Xba*I-restriction pattern of chromosomal DNA from *A.thiooxidans* strain TTO from reactor 1 (1) and reactor 3 (2), size marker (3). Conditions of electrophoresis: voltage, 120 V, pulse time, 25 s, duration, 44 h.

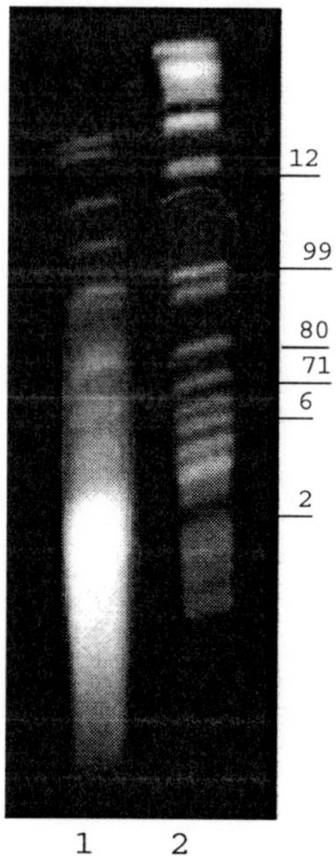

Fig. 6. *Not*I-restriction pattern of chromosomal DNA from *Sulfobacillus* sp. strain Kr1. Conditions of electrophoresis are as in the legend to Fig. 5.

Obviously, the aboriginal strains of chemolithotrophic bacteria are active in the bioleaching processes, but their functional activity is different in different reactors, depending on changes in the mineral composition of the solid phase and physicochemical conditions in the pulp.

Distribution of Chemolithotrophic Bacteria in the Pulp

In the pulp, which is a complex heterogeneous system, bacteria are distributed in its liquid and solid phases. However, no quantitative data on the distribution of chemolithotrophs in dense pulps processed for a longed time under continuous conditions are available. Moreover, the roles of attached and floating cells in bioleaching of complex products have not been understood yet.

Properties of the Pulp Phases

From reactors 1 and 4, we took 500-ml pulp samples differing in the oxidation degree of sulfide minerals, as well as in physicochemical and chemical properties. The pulp samples was allowed to settle in 1-l cylinders for 12 h. As result of precipitation, three phases appeared, namely the liquid phase and two solid phases, the top and bottom ones (Table 3), which differed in their volumes, dry weights, colors, and densities. The top solid phase (TSP) mainly contained slimes, jarosites, hydroxides, and fine fractions of unoxidized sulfide minerals, whereas the bottom solid phase (BSP) was composed of more coarse sulfide mineral particles. The data of chemical analyses of the solid phases are summarized in Table 4. As can be seen from these data, TSP in the pulp of both reactors was characterized by a relatively low content of sulfide minerals and sulfur. There were significant differences between the bottom solid phases of the pulp in the two reactors; thus, the BSP from reactor 4 was characterized by smaller volume and more profoundly oxidized sulfide minerals than in the BSP from reactor 1.

Table 3. Characteristics of the pulp phases

Parameters	Reactor 1	Reactor 4
Pulp volume, ml	500	500
Liquid phase volume, ml	420	435
TSP volume, ml	50	50
BSP volume, ml	30	15
Air-dry TSP weight, g	50.46	61.46
Air-dry BSP weight, g	31.37	20.73
Air-dry solid phase weight, g	81.83	82.19

Table 4. Mineral composition of the pulp solid phases (%).

Components	Reactor 1		Reactor 4	
	TSP	BSP	TSP	BSP
Total arsenic	1.15	1.88	0.96	0.83
Sulfide arsenic	0.5	1.48	0.4	0.57
Total antimony	0.49	0.56	0.38	0.32
Sulfide antimony	0.35	0.45	0.34	0.22
Total sulfur	10.89	18.08	10.21	10.61
Sulfide sulfur	4.43	16.18	3.49	8.33
Elemental sulfur	1.28	3.62	0.93	2.75
Total iron	24.11	18.99	23.26	13.01

Distribution of Chemolithotrophic Microorganisms in Pulp Phases

The total number of bacterial cells in various phases of the processed pulp was determined to estimate the distribution pattern of chemolithotrophic bacteria (Table 5).

Table 5. Distribution of bacteria in the pulp from reactors 1 and 4.

Analyzed pulp portion	Bacterial cell number	
	Reactor 1	Reactor 4
1-ml aliquot of liquid phase. Total liquid phase, 420 (435) ml. (See Table 3).	$4.0x10^9$ $16.8x10^{11}$	$4.6x10^9$ $20.01x10^{11}$
Water wash from TSP	$0.35x10^{11}$	$1.25x10^{11}$
0.0004 N NaOH wash from TSP	$0.028x10^{11}$	$0.027x10^{11}$
Water wash from BSP	$0.048x10^{11}$	$0.11x10^{11}$
0.0004 N NaOH wash from BSP	$0.0045x10^{11}$	$0.0047x10^{11}$
Total solid phase	$0.43x10^{11}$	$1.39x10^{11}$
Total pulp, 500 ml	$17.23x10^{11}$	$21.40x10^{11}$
1-g aliquot of solid phase	$5.25x10^8$	$1.69x10^9$
	% of cells	
Washed-off with water from TSP	81.4	89.9
Washed-off with 0.0004 N NaOH from TSP	6.51	1.94
Washed-off with water from BSP	11.16	7.9
Washed-off with 0.0004 N NaOH from BSP	1.05	0.34
In the liquid phase (% of total)	97.5	93.5
In the solid phase (% of total)	2.5	6.5

Noteworthy, the major part of chemolithotrophic bacteria was present in the liquid phase of pulp, accounting for 97.5% and 93.5% of the total cell number in reactors 1 and 4, respectively. The bacterial cell numbers per 1 ml of the liquid phase from reactors 1 and 4 were similar and comprised respectively $4.0x10^9$ ml^{-1} and $4.6x10^9$ ml^{-1}. After washing of the solid pulp phase with distilled water, the major part (92.6 - 97.8%) of bacterial cells was transferred into the aqueous bulk; subsequent washing of the solid phase with 0.0004 N NaOH solution released 7.6% and 2.3% of total cells from the pulp of reactors 1 and 4, respectively. Taken together, these data indicated that the studied bacteria were loosely attached to mineral particles.

The total bacterial numbers per 1 g of air-dry solid phase from reactors 4 was 3.2-fold higher than that in the solid phase of the pulp in reactor 1. Since the most part of bacterial cells was washed off with water from the solid phase, it is reasonable to conclude that they were mechanically trapped by hydroxide, jarosite, and slurry particles. As to the cells tightly bound to BSP in reactor 1, it is possible that during the continuous processing of the concentrate they may be transferred to reactors 3 and 4 together with BSP.

Thus, our studies demonstrated that major part (93.5-97.5%) of chemolithotrophic bacteria occurred in the liquid phase of the processed pulp. The number of bacterial cells that were firmly attached to particles of TSP and BSP amounted to 3.2×10^9 ml^{-1} and 3.17×10^9 in reactors 1 and 4, respectively. The significance of adsorbed cells, constituting a considerable biomass, should not be underestimated. Evidently, the bacterial cells attached to sulfide minerals play a role in intensification of the bioelectrochemical processes in the pulp as a whole and in implementation of solid-phase diffusion-metasomatic reactions (e.g., S^{2-}/S^0 oxidation) on the surface of sulfide minerals. Adsorbed bacteria influence the electrode potential of sulfide minerals and sulfur [17, 18], their charge [19-21], electroconductivity, resistance, type of conductivity, etc. [12, 16, 17]. In the liquid phase, the activity of chemolithotrophic bacteria resulted in the oxidation of Fe^{2+} and soluble sulfur-containing compounds, thus contributing to the high redox potential in the pulp and electroconductivity of the medium (electrolyte) [3, 12-18]. Such a heterogeneous and non-equilibrated system provides for conditions favorable for profound oxidation of sulfide minerals and gold liberation. Evidently, the solid and liquid pulp phases represent themselves a combined bioelectrochemical system, composed of the electrolyte medium and oxidized substrates.

In conclusion, it should be emphasized that our experiments demonstrated a most important role of the community of mesophilic and moderately thermophilic bacteria in the oxidation of sulfide minerals in reactors during processing of pyrrhotite gold-arsenic concentrate in a pilot bioleaching plant. Under mesophilic conditions at 30^0C, the *A. ferrooxidans - A. thiooxidans* community was predominant in the oxidative processes. However, with an increase in the temperature to 40^0C and higher due to exothermal oxidative processes, as it has already been observed under industrial conditions, the association of thermophilic bacteria will dominate and determine the successful implementation of the bioleaching technology.

ACKNOWLEDGMENTS

This work was supported by grants nos. 01-04-48270 and 03-04-49294 from the Russian Foundation for Basic Research, a grant within the scope of the Federal Central Scientific and Technical Program "Technology of the Living Systems" (no. 43.073.1.1.2515) , and a grant from the "The Leading Science Schools" no. 1742.2003.4.

REFERENCES

1. C.L. Brierley, A.P. Briggs: In: *Mineral Processing Plant Design, Practice and Control Processing*. Mular A.L., D.H. Halbe and D.J. Barret (Eds.). Soc. Metallurg. Eng. Littleton, Colorado. 1540 (2002).
2. D.E. Rawlings, H. Tributsch, G.S. Hansford: *Microbiol. (UK),* **145** (1), 5 (1999).
3. G.I. Karavaiko et al. (Eds): *Biogeotechnology of Metals*. Manual. CIP GKNT, 1989. 375 p.
4. J. Deley, H. Cattoir, A. Reynaerts: *Eur. J. Biochem,* **12** (1), 133 (1970).
5. G.I. Karavaiko, T.P. Tourova, T.F. Kondrat'eva, A.M. Lysenko, T.V. Kolganova, S.N. Ageeva, L.N. Muntyan, T.A. Pivovarova: *Intern J. Syst. and Evol. Microbiol.,* **53** (1), 113 (2003).
6. T.F. Kondrat'eva, L.N. Muntyan, G.I. Karavaiko: *Microbiol. (UK),* **141** (5), 1157 (1995).
7. A.A. Reznikov, E.P. Mulikovskaja, I.Yu. Sokolov: *Methods of analysis of the natural waters*. NEDRA, Moscow. 137 (1970). (in Russian).
8. O.V. Golyshina, T.A. Pivovarova, G.I. Karavaiko, T.F. Kondrat'eva, E.R.B. Moore, W-R. Abraham, H. Lündsdorf, K.N. Timmis, M.M. Yakimov, P.N. Golyshin: *Intern. J. System and Evol. Microbiol.,* **50** (3), 997 (2000).
9. T.A. Pivovarova, T.F. Kondrat'eva, S.G. Batrakov, E.E. Esipov, V.I. Sheichenko, S.A. Bykova, A.M. Lysenko, G.I. Karavaiko: *Microbiol.,* **71** (6), 809 (2002). (in Russian).
10. J. Romero, R.T. Espejo, C. Yanez, M. Vasquez, E.R.B. Moore: *Research in Microbiology,* **154** (235), 353 (2002).
11. M. Vasquez, R.T. Espejo: *Appl. Environ. Microbiol.* **63** (2), 332 (1997).
12. G.I. Karavaiko, L.S. Smolskaja, O.K. Golyshina, M.A. Jagovkina, E.Y. Egorova: *Fuel Proc. Technol.,* **40** (2+3), 152 (1994).
13. G.I. Karavaiko, T.V. Kovalenko, R.S. Golovacheva: In. *Proc. Intern. Conf. on Use of Microorganisms in Hydrometallurgy*. Dec. (Hung.): Pecs. 95 (1980).
14. B. Johnson, P.B. Nicolau, N. Okibe, A. Yahua, K.B. Hallberg: In: *Biohydrometallurgy: Fundamentals, Technol. and Sustainable. Development. Part A*. V.S.T. Ciminelli, O. Garcia Ir. (Eds). Elsevier, Amsterdam, London etc, 461 (2001).
15. N.S. Vardanjan: *Biotechnology* (2004). (in Russian) (in press).
16. T.F. Kondrat'eva, O.V. Tupikina, S.N. Ageeva, V.D. Samorukova: In. Proceedings II Moscow Intern. Congr. Biotechnology: *State of the art and Prospects of Development* (in press).
17. L.K. Yakhontova, V.M. Sergeev, G.I. Karavaiko, V.S. Sukhantseva: In: *Ecology and Geochemical Activity of Microorganisms*. Pushchino, 99 (1976) (in Russian).
18. L.K. Yakhontova, A.P. Grudev: *Hypergenesis Zone of Ore Deposits*. Moscow State University, 1978. 229 p. (in Russian).
19. R.C. Blake II, E.A. Shute, G.T. Howard: *Appl. and Environ. Microbiol.*, **60** (9), 3349 (1994).
20. R.C. Blace II, M.M. Lyles, R.C. Simmons: In. *Biohydromet. Processing*. T. Vargas, C.A. Jarez, J.V. Wiertz, H. Toledo (Eds.). University of Chile, 1,13 (1995).
21. G.I. Karavaiko, T.V. Sakhno, V.M. Emelyanov, T.A. Pivovarova: In. *Biohydromet. Processing*. T. Vargas, C.A. Jarez, J.V. Wiertz, H. Toledo (Eds.) University of Chile, 1, 23 (1995).

Chapter 19

Viability and Metabolic Activity of Microorganisms under Conditions Emulating Natural Habitats

O. V. Heidebrecht, O. V. Shelemekh*[*], *V. K. Plakunov, V. G. Arzumanyan and S. S. Belyaev
Institute of Microbiology, Russian Academy of Sciences,
Moscow, 117312 Russia, fax (095) 135-65-30

Abstract

The biochemical mechanisms were studied that determine different reactions of bacteria and yeasts of different genera to two simultaneously imposed stressors, hypoxia and osmotic shock. For *Candida lipolytica*, *Candida rhagii*, *Debaryomyces hansenii*, *Rhodotorula aurantiaca Rhodococcus erythropolis*, *Shewanella* sp CN32, and *Halobacterium salinarum* S9 these two stressors were antagonistic, which resulted in stimulation (not suppression) of the growth of these microorganisms by NaCl (in a wide range of concentrations) under microaerobic conditions. The reaction of *Malassezia* sp. was different: the degree of halotolerance of this microorganism was lower under microaerobic conditions. These differences were found to be determined, above all, by the induction of a salt-resistant respiratory system (oxidase) in most organisms under study, which could not be induced in *Malassezia* sp. In addition, the synthesis of catalase was enhanced in all the yeasts except for Malassezia, which provided for neutralization of the active forms of oxygen accumulating as a result of inhibition of other protective enzymes by salt. Also, the effect of simultaneous influence of enhanced proton concentrations and hypoosmotic shock on extremely halophilic *Halobacterium salinarum* was studied. Acidification of the growth medium, i.e., an increase in proton concentration, was found to stabilize growth of H. salinarum at low salt concentrations. The implications of these phenomena for the ecophysiology of microorganisms and enhancement of oil production are discussed.

---

[*] Corresponding author: omaslova@yandex.ru

Key words: multiple stress, hypoxia, osmotic stress, halophilic microorganisms, osmoprotection, cross protection

Generally, optimal cultivation parameters determined for microorganisms in a laboratory differ from what they face in different environments. In natural habitats microorganisms are exposed to multiple stressors. Nevertheless, they retain their metabolic activity and reproductive potential, which may be ensured by the phenomenon of antagonism between different stressors when applied simultaneously.

Recently, high emphasis has been placed on the influence of various stressors on the growth and metabolism of microorganisms. Multiple stress simulating the conditions of natural habitats is of great interest. Most researchers choose *Escherichia coli* as the object of such studies and enhanced temperatures and salt concentrations as stressors. However, studies have been conducted with other microorganisms (mainly prokaryotes) and stressors. For example, the biochemical effects of high salt concentrations and enhanced temperatures have been studied in *Streptomyces* species [1]. *Lactococcus lactis* was found to gain resistance to some stressors after incubation in subacid medium [2]. Sometimes in natural habitats (particularly, in oil fields), aerobic microorganisms concurrently have to cope with a deficiency in oxygen (hypoxia) and enhanced salt concentrations. However, no data are available on the effects caused by a combination of these stressors in microorganisms. On the other hand, the same salt conditions may appear hypoosmotic for halobacteria, which strictly depend on the presence of salt in the environment, with optimal NaCl concentration of 20%. Halobacterial cells swell and lyse at less than 10-15% NaCl. However, halobacteria are frequently isolated from saline soils [3] and arid soils [4] (where salinity fluctuations are possible as results of floods and precipitation), stratal waters of oil fields (where there are areas with a salinity of less than 15%) and even sea water with salt content below 5% [5].

We studied the effects produced by the simultaneous influence of hypoxia (microaerobic conditions) and high salt concentrations (hyperosmotic shock) on a wide range of microorganisms: oil-oxidizing slightly halophilic bacteria *Rhodococcus erythropolis* and *Shewanella* sp CN32, extremely halophilic *Halobacterium salinarum* S9, slightly halophilic yeasts *Debaryomyces hansenii, Rhodotorula aurantiaca, Malassezia* sp., moderately halophilic yeast *Candida rhagii*, and halotolerant yeast *Candida lipolytica*. Also, we studied simultaneous influence of hypoosmotic conditions and enhanced concentrations of protons on the growth of *Halobacterium salinarum* S9.

RESULTS AND DISCUSSION

Under aerobic conditions, *Shewanella* sp CN32 and *Rh. erythropolis* should be regarded as slightly halophilic (optimal NaCl concentrations for their growth ranged from 1 to 3%). However, under microaerobic conditions maximal growth rate was observed at 5-6% NaCl, which is characteristic of moderately halophilic microorganisms (Fig. 1, 2).

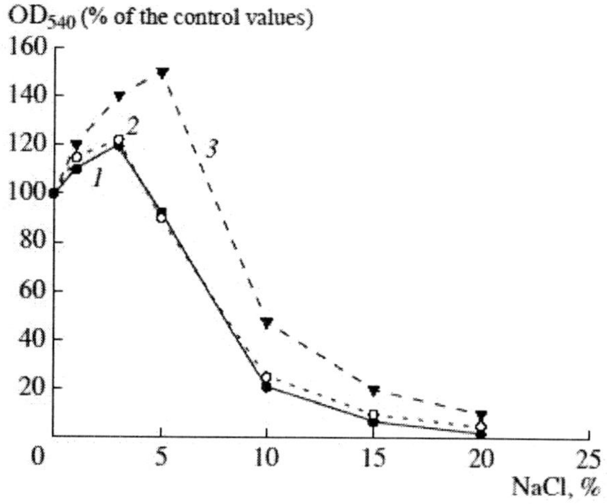

Fig. 1. Growth of *Shewanella* sp. CN32 at various NaCl concentrations and aeration levels: (1) aerobic conditions with agitation; (2) stationary aerobic conditions; (3) microaerobic conditions. Control cultures were grown under the specified conditions but without extra NaCl added.

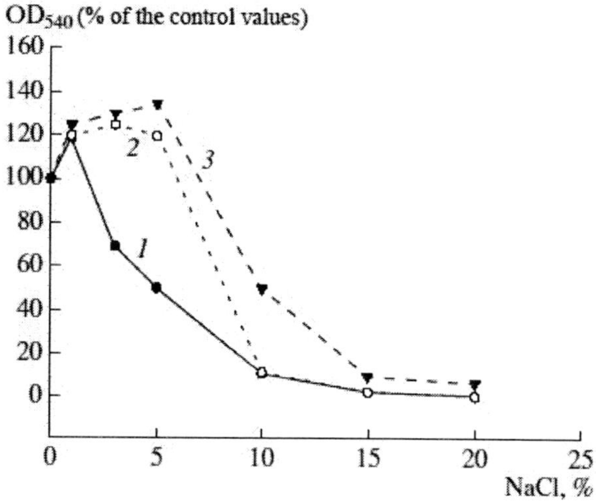

Fig. 2. Growth of *Rhodococcus erythropolis* at various NaCl concentrations and aeration levels: (1) aerobic conditions with agitation; (2) stationary aerobic conditions; (3) microaerobic conditions.

Extremely halophilic *H. salinarum* can grow in media containing near-saturation NaCl concentrations (to 34%). However, under microaerobic conditions a certain increase in the degree of halophily of the culture was observed. Although the NaCl concentration of 20% remained optimal, the growth rate at supraoptimal salt concentrations was higher than that under aerobic conditions (Fig. 3).

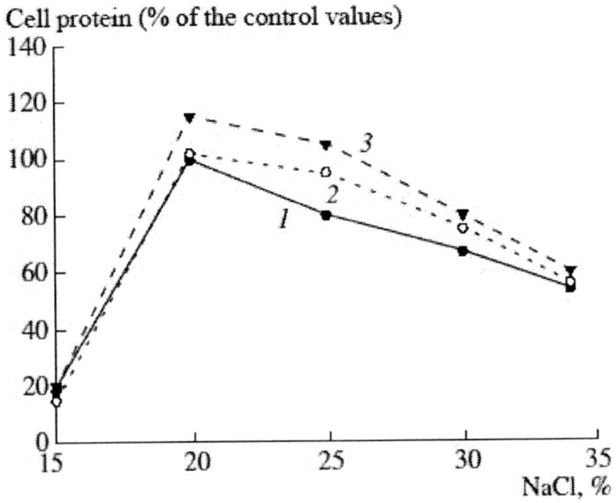

Fig. 3. Growth of *Halobacterium salinarum* S9 at various NaCl concentrations and aeration levels: (*1*) aerobic conditions with agitation; (*2*) aerobic conditions without agitation (stationary); (*3*) microaerobic conditions. The control culture was grown at a NaCl concentration of 20%.

For yeast cultures, the degree of halotolerance depended on several factors. First, with the time of cultivation, the degree of halotolerance became higher. For example, for aerobically cultivated *C. lipolytica*, the NaCl concentration that reduced the growth rate by 50% (ID_{50}) was 5–12% on the fourth day of growth and 14–15% on the ninth day. Second, the degree of halotolerance depended on the efficiency of energy and carbon sources. For *C. lipolytica* grown on glucose medium under aerobic conditions, ID_{50} (on the fourth day) was 5–7.5%; on glycerol-containing medium, it was 7.5–10%; and on YPD medium, 12%. Above all, the degree of halotolerance was influenced by microaerobic conditions. The data are given in Fig. 4 - 7. The growth rate of all cultures decreased significantly under microaerobic conditions in the range of NaCl concentrations to which, under aerobic conditions, all the yeasts were tolerant (0 to 10–12%). However, at higher salt concentrations, the difference between the growth rates under aerobic and microaerobic conditions was minor (*Malassezia* sp. was an exception); and, at 15–20% NaCl, the growth rate was sometimes higher under microaerobic than under aerobic conditions. This is evident if the relative growth rates are compared (the culture grown without extra NaCl added taken as control). Although the biomass gain under microaerobic conditions was at least one order less, the range of optimal NaCl concentrations and/or growth maximum differed from those under aerobic conditions. The extension of the range of optimal NaCl concentrations, and a sharp maximum of the growth rate at 8-10% NaCl became evident for *C. lipolytica*, i.e. the culture became moderately halophilic. For *Rh. aurantiaca,* higher growth rates at enhanced salt concentrations were observed. The growth of *D. hansenii* under microaerobic conditions was characterized by an increase of NaCl concentrations at which maximum biomass gain was observed (the degree of halophily rose); whereas *C. rhagii* behaved as an extreme halotolerant (the extension of the range of optimal NaCl concentrations was evident). For *Malassezia* sp., on the contrary, growth was observed in a narrower range of NaCl concentrations, i.e. the degree of halotolerance decreased.

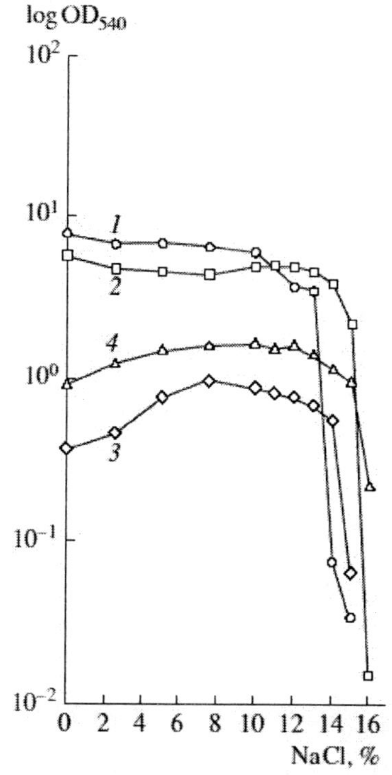

Fig. 4. Dependence of the degree of halotolerance of *Candida lipolytica* on aeration conditions: (1, 2) aerobic conditions; (3, 4) microaerobic conditions; (1, 3) fourth day; (2, 4) ninth day.

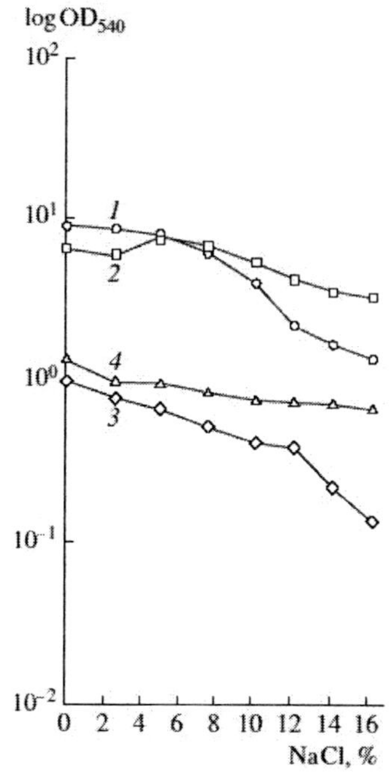

Fig. 5. Dependence of the degree of halotolerance of *Rhodotorula aurantiaca* on aeration conditions: (1, 2) aerobic conditions; (3, 4) microaerobic conditions; (1, 3) fourth day; (2, 4) ninth day.

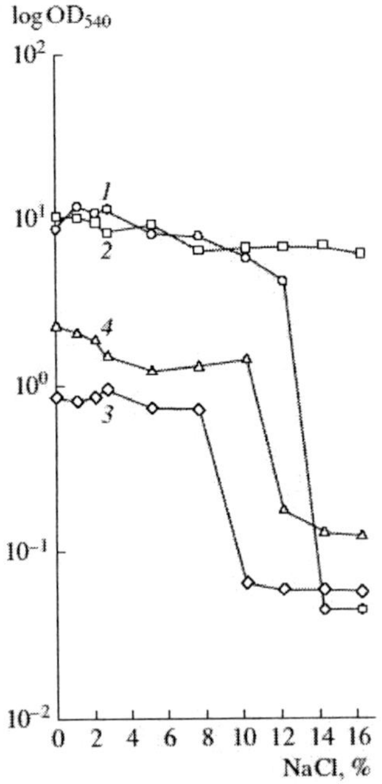

Fig. 6. Dependence of the degree of halotolerance of *Malassezia* sp. on aeration conditions: (1, 2) aerobic conditions; (3, 4) microaerobic conditions; (1, 3) fourth day; (2, 4) ninth day.

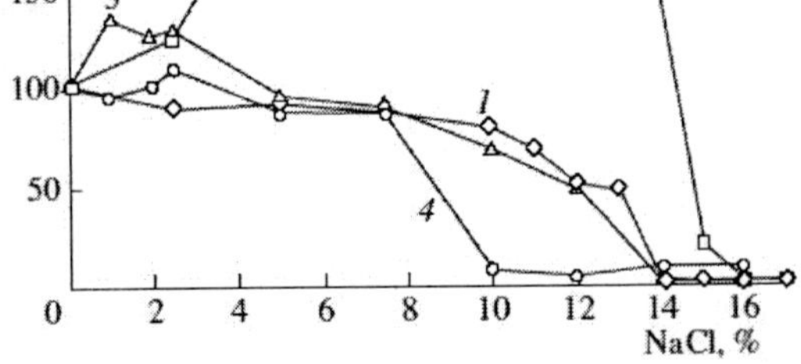

Fig. 7. Relative growth rates of *Candida lipolytica* and *Malassezia* sp. (cultivated for four days): (1, 3) aerobic conditions; (2, 4) microaerobic conditions; (1, 2) *Candida lipolytica*; (3, 4) *Malassezia* sp.

Thus, the stressors under study were antagonistic for all the bacterial and yeast cultures examined except *Malassezia* sp. The reasons of these differences were studied in more detail. An increase in halophily (halotolerance) degree under microaerobic conditions is accompanied by multiple changes in metabolic processes. Enhanced salt concentrations suppress respiration and inhibit enzymes which protect cells from active forms of oxygen (superoxide dismutase and catalase). However, the level of oxygen in a cultivation medium remains high (the decrease in solubility in the presence of salt does not exceed 10-15%). Thus, reduction in oxygen level (microaerobic conditions) compensates for the deficiency in protective mechanisms and allows growth at higher salt concentrations. Also, the intracellular levels of osmoprotectors increase due to the retardation of oxidation processes.

We observed a considerable decrease in the respiration rate of cultures grown in media without extra NaCl if 15% NaCl was added to the incubation mixture. The respiration rate of *C. lipolytica, Rh. aurantiaca, C. rhagii, D. hansenii* and *Malassezia* sp. decreased by 50-80%. *Rh. erythropolis* and *Shewanella* sp. CN32 exhibited a decrease of 80-90%. However, if cultures (except *Malassezia* sp.) were grown at 15% NaCl, the addition of NaCl to the incubation mixture resulted not in inhibition but in stimulation of respiration. The strongest stimulation was observed in *C. lipolytica* when NaCl was added to the incubation mixture at the same concentration in which it was present in the growth medium (15%). This can be explained by the induction of an alternative, salt-resistant respiratory chain in these microorganisms when they are grown at high salt concentrations, whereas *Malassezia* sp. is incapable of such induction. This assumption is in accordance with the available data whereby the organization of the respiratory chain in *C. lipolytica* changes at high NaCl concentrations; in particular, the activity of succinate oxidase increases [6].

The enhanced salt concentrations drastically inhibited catalase activity in all the bacteria and yeasts studied. This was especially pronounced for *Malassezia* sp., in which the activity dropped by 80%. For yeasts, potential catalase activity was also determined (catalase activity in cells grown at 15% NaCl measured in an incubation medium without salt). For *C. lipolytica, C. rhagii, D. hansenii* and *Rh. aurantiaca*, potential catalase activity was much higher than that of control variant; whereas for *Malassezia* sp. it comprised only 13% of the control level (catalase activity in cells grown without extra NaCl and determined in an incubation mixture also containing no salt was taken for control). Therefore, we concluded that in *C. lipolytica, C. rhagii, D. hansenii* and *Rh. aurantiaca* exposed to osmotic shock, along with the inhibition of catalase activity, the synthesis of catalase (of the same or a new form of this enzyme) is induced. Thus, under microaerobic conditions, these microorganisms can efficiently neutralize active forms of oxygen and, as a result, exhibit an increased degree of halophily (halotolerance). On the contrary, in *Malassezia* sp. exposed to osmotic shock, catalase synthesis is not induced, and active forms of oxygen cannot be neutralized by the available quantities of the enzyme, even if oxygen partial pressure is low. As a result, *Malassezia* sp. becomes less halotolerant under microaerobic conditions.

General conclusions about the relations of the two stressors under study can be formulated as follows. The presence of salt (in addition to osmotic effects per se) causes suppression of respiration and inhibition of enzymes protecting the cell from active forms of oxygen. Thus, in the presence of salt, cells undergo oxidative stress. The stressors studied are antagonistic for the microorganisms in which a salt-resistant respiratory system and synthesis of extra quantities of salt-resistant protective enzymes are induced (for example, *C. lipolytica*) and additive for the organisms incapable of such induction (*Malassezia* sp.).

Stress response is obviously regulated by multiple mechanisms. One of the possible responses of microorganisms to simultaneous osmotic shock and hypoxia is an increase in the intracellular concentrations of osmoprotectors (primarily amino acids) under microaerobic conditions, which may be due to the deceleration of the processes of oxidation of these compounds. This is in accordance with the data of other researchers, who revealed the protective effects of some metabolites, including those functioning as osmoprotectors, under the influence (not simultaneous but separate) of some stress factors. For example, mutant strains of *Schizosaccharomyces pombe* with the impaired trehalase gene, which accumulate trehalose to a greater extent than wild-type strains, are more resistant to heat shock and osmotic shock [7]. Mutant cells of *Saccharomyces cerevisiae* containing a plasmid coding for mannitol-1-phosphate dehydrogenase accumulate mannitol and are more resistant to oxidative and osmotic stresses [8]. We estimated the levels of intracellular organic solutes at enhanced salt concentrations under both aerobic and microaerobic conditions in one bacterial (*Rh. erythropolis*) and one yeast (*C. lipolytica*) cultures. Under microaerobic conditions, cells of *Rh. erythropolis* and *C. lipolytica* contained 1.3-1.5 times higher levels of intracellular organic solutes at enhanced salt concentrations than under aerobic conditions. The most pronounced increase was observed in the concentrations of amino acids, which serve as osmoprotectors in slightly and moderately halophilic microorganisms (in rhodococci, glutamic acid and serine; in *C. lipolytica*, proline and alanine mainly). This may contribute to the effect of a halophily (halotolerance) increase under microaerobic conditions.

Thus, hypoxia and high salt concentrations may display an antagonistic mode of interaction when applied in combination. This phenomenon is widespread among both prokaryotic and eukaryotic microorganisms in a wide range of salt concentrations.

We found that the cells of *H. salinarum*, when subjected to hypoosmotic stress, are stabilized by enhanced proton concentrations. Earlier, we described a phenomenon of halodisadaptation (termed by analogy with haloadaptation, adaptation to a high-salinity environment) occurring among halophilic bacteria. Halodisadaptation is a complex of processes allowing bacteria to adapt to low-salinity conditions [9]. In halobacteria, products of cell osmotic lysis are among the most important disadaptation factors, acting as signals of hypoosmotic stress and participating in osmotic stabilization [10]. Short-term stabilization of halobacterial cells can also be achieved by replacement of sodium cations, which stabilize bacterial cell, with protons. Proton-stabilized cells endure even a wash with distilled water [11]. In this work, we studied a similar effect in growing cultures.

As stated above, the optimal for growth NaCl concentration for *H. salinarum* was found to be 20%. At 20% NaCl, the optimal pH values ranged from 7.0 to 7.2. However, at 15% NaCl, the growth rate decreased dramatically at pH 7.0 and especially at pH 8.0 (five- to sixfold), whereas at pH 5.5-6.0 no differences in growth were observed in comparison with 20% NaCl (Fig. 8). This effect becomes even more obvious when comparing relative growth rates expressed as percentages of the growth rate at the same salinity but at the potimal pH 7.0. At pH 6.0, a sharp maximum of the relative growth rate at 15% NaCl is evident, whereas at other salt concentrations, the relative growth rate peaks at 7.0 (Fig. 9). Therefore, acidification of the growth medium, i.e., an increase in proton concentration, stabilizes growth of H. salinarum at low salt concentrations. This effect does not seem to require acidification of large volumes of extracellular fluids, because halobacteria are capable of creating a local proton gradient within their surface structures, with the outer cell wall preventing proton leaching from the cell.

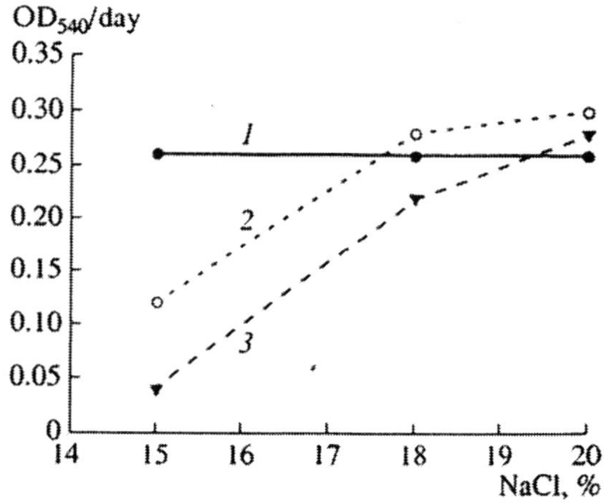

Fig 8. Effect of NaCl concentration on the growth rate of *Halobacterium salinarum* S9 at different pH values: (1) pH 6.0; (2) pH 7.0; (3) pH 8.0.

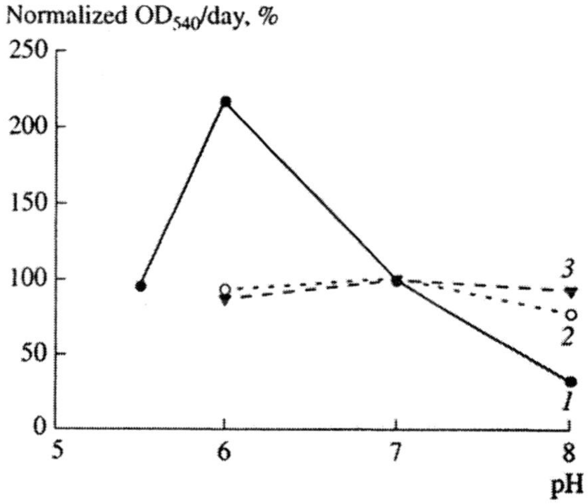

Fig. 9. Effect of pH on the relative growth rate of *Halobacterium salinarum* S9 at different NaCl concentrations: (1) 15%; (2) 18%; (3) 20%.

The described phenomena are important for estimation of metabolic activity of microorganisms in their natural environments. The combination of hypoxia and high salt concentrations quite frequently occur in nature. In particularly, this is typical of oil fields. However, slightly halophilic aerobic microorganisms are quite often isolated from such ecotopes. This can be explained by the effect of cross protection, which is significant for survival under multiple stresses. Hypoxia reduces the inhibitory effects of high salt concentration, allowing growth and population maintenance. Halobacteria, in their turn, often experience hypoosmotic stress in their habitats. In oil fields, for example, stratal waters may contain areas with salinity of less than 150g/l. Slightly acidic conditions resulting from the accumulation of fatty acids and oxygen deficiency should ensure the survival of these bacteria in a wide range of NaCl concentrations. The data obtained can be used for the improvement of microbial methods of petroleum production.

References

1. G. Malin, A. Lapidot: Induction of Synthesis of Tetrahydropyrimidine Derivatives in *Streptomyces* Strains and Their Effect on *Escherichia coli* in Response to Osmotic and Heat Stress. *J. Bacteriol.,* **178**, 385 (1996).
2. E. O'sullivan,S. Condon: Intracellular pH Is a Major Factor in the Induction of Tolerance to Acid and Other Stresses in *Lactococcus lactis. Appl. Environ. Microbiol.,* **63**, 4210 (1997).
3. S. Zvyagintseva, A. L. Tarasov: Extremely Halophilic Bacteria from Saline Soils. *Mikrobiologia,* **56** (5), 839 (1987).
4. S. Kulichevskaya, I. S. Zvyagintseva, A. L. Tarasov, V. K. Plakunov: Extremely Halophilic Archaebacteria from Some Saline Ecotopes. *Mikrobiologia,* 1992, **61** (1), 70 (1992).
5. F. Rodriges-Valera, F. Ruiz-Berraquero, A. Ramos-Cormnzana: Isolation of Extreme Halophiles from Seawater. *Appl. Environ. Microbiol.,* **38**, 164 (1979).
6. E. N. Andreishcheva, M. I. Soares, R. A. Zvyagil'skaya: Energy Metabolism of the Yeast *Candida* (*Yarrowia*) *lipolytica* under Normal Conditions and under Salt Stress. *Fiziol. Rast.,* **44** (5), 658 (1997).
7. J. Cansado, T. Soto, J. Fernandes, J. Vincentesoler, M. Gacto: Characterization of Mutants Devoid of Neutral Trehalase Activity in the Fission Yeast *Schizosaccharomyces pombe*: Partial Protection from Heat Shock and High-Salt Stress. *J. Bacteriol.,* **180**, 1342 (1998).
8. V. Chaturvedi, A. Bartiss, B. Wong: Expression of Bacterial *mtlD* in *Saccharomyces cerevisiae* Results in Mannitol Synthesis and Protects a Glycerol-Defective Mutant from High-Salt and Oxidative Stress. *J. Bacteriol.,* **179**, 157 (1997).
9. N. I. Matveeva, Yu. A. Nikolaev, V. K. Plakunov: Dependence of the Amino Acid Transport into Cells of Halophilic and Halotolerant Bacteria on the Concentration of NaCl and Osmolarity of the Medium. *Mikrobiologia,* **59** (1), 5 (1990).
10. M. V. Kokoeva, V. K. Plakunov: The Possibility of Modification of the Osmosensitivity of Extremely Halophilic Archaebacteria. *Mikrobiologia,* **62** (5), 825 (1993).
11. V. K. Plakunov, M. V. Kokoeva: Osmostabilization of Halobacterial Cells and Obtaining of Their Dry Biomass Free from Salts. *Mikrobiologia,* **63** (4), 597 (1994).

Chapter 20

THE DYNAMICS OF MICROBIOLOGICAL PROCESSES ASSOCIATED WITH ATTEMPTS TO ACTIVATE MICROBIAL COMMUNITIES IN OIL-BEARING STRATA

I. A. Borzenkov* [*], *I. A. Tarasov, E. I. Milekhina and S. S. Belyaev
Institute of Microbiology, Russian Academy of Sciences,
Moscow, 117312 Russia

ABSTRACT

Experiments that were performed in this work showed a feasibility of activating microbiological processes in the near-bottom zones of injection wells of oil-bearing strata into which freshwater was injected. It was found that a rather long period of activation of sulfate-reducing and methanogenic bacteria in the near-bottom zone of an injection well was followed by a period of a gradual decrease in microbial activity to a background level.

Key words: oil-bearing stratum, improved oil recovery, biogeotechnology, microflora of oil deposits, sulfate reduction, methanogenesis

Current methods of oil production allow the recovery of no more than 50% of the oil reserve, which poses the problem of improvement of oil recovery. Nowadays, a large number of oil deposits are exploited by using waterflood. In this case, water circulation for many years results in the waterflooded oil-bearing strata containing different microorganisms and certain chemical compounds. The study of the resulting technogenic microbial community is of both scientific and practical interest.

---

[*] Corresponding author. E-mail: borzenkov@rbcmail.ru. Fax: (095) 135-65-19.

MATERIALS AND METHODS

Microbiological studies were carried out at different areas of the Romashkino Oil Field, whose oil-bearing Devonian and Bobrikovsky sedimentary rocks are represented by fine quartz sandstones with argillite or aleurolite intercalations. The in situ temperatures of the Devonian and Bobrikovsky rocks were 30–35°C and 25°C, respectively. The original oil field water was a chloride calcium brine with a total mineralization of up to 320 g/l. To maintain the formation pressure at a sufficient level, the oil field is presently injected with fresh water or highly salinity wastewater. Samples for investigation were taken from wells as described earlier [9].

Microorganisms were enumerated by the serial tenfold dilution method using selective media. Saprotrophic microorganisms were detected with the use of GPY (glucose–peptone–yeast extract) agar medium [2]. Hydrocarbon-oxidizing microorganisms were detected using a modified version of Raymond medium that was supplemented either with Devonian crude oil from the Romashkino Oil Field or with liquid paraffin (Parex) as the organic substrate [3, 4].

Anaerobic microorganisms were counted by the serial tenfold dilution method using appropriate liquid media and employing an anaerobic technique. Aliquots (0.5 ml) of the formation liquid samples were placed in Hungate or Balch tubes that contained 4.5 ml of an appropriate medium, and the tubes were incubated at 30°C for a month. Sulfate-reducing bacteria were counted using liquid Postgate B medium and estimating the formation of either iron sulfide [3, 5] or hydrogen sulfide [6]. Different groups of methanogenic bacteria were distinguished by their substrate specificity and the degree of mineralization of the growth medium. The growth of sulfate-reducing bacteria was inhibited by adding 5 g/l Na_2MoO_4 to the incubation medium [7]. The basal media were PPBM and HPM (halophile medium) [8–10], which were supplemented with one of the following carbon and energy sources: (1) a $CO_2 + H_2$ (20 : 80) gas mixture; (2) 5 g/l acetate and a $CO_2 + N_2$ (20 : 80) gas mixture; (3) methanol and trimethylamine (MTMA) at a concentration of 5 g/l each plus the $CO_2 + N_2$ gas mixture. The growth of methanogens was estimated from the increase in the methane content of the headspace gas. Acetogenic bacteria were enumerated by using PPBM [8] with the $CO_2 + H_2$ gas mixture. To inhibit the growth of methanogens, the medium was supplemented with 2-bromoethanesulfonic acid (an inhibitor of methanogenesis) [7]. The growth of acetogens was estimated from the increase in the acetate content of the incubation medium. Acetate and other fatty acids were analyzed by gas–liquid chromatography after their extraction and the evaporation of the extract [11]. The rates of bacterial sulfate reduction and methanogenesis from CO_2 and acetate were determined using the radiolabeled compounds $NaH^{14}CO_3$, $^{14}CH_3COONa$, and $Na_2{}^{35}SO_4$ [3, 12]. The radioactivity of the hydrogen sulfide and methane produced was determined with the use of a RackBeta liquid scintillation counter. The stable carbon isotopes of carbonates dissolved in the medium and those of methane present in the headspace gas were analyzed using an MI-1201 mass spectrometer [11, 13]. The carbonates were preliminarily precipitated from the stratal water samples by adding a saturated solution of $Ba(OH)_2$, whereas the methane was purified on a preparative column with Porapak Q and then burnt to CO_2.

RESULTS AND DISCUSSION

Oil fields exploited in Tatarstan are flooded with surface freshwater or highly mineralized wastewater, which results in the stratal water possessing a variable mineralization. Table 1 summarizes the biogeochemical characteristics of the stratal water versus its mineralization. It can be seen that the activity of microbiological processes in the stratal water tended to diminish as its mineralization increased.

Table 1. Change of certain biogeochemical characteristics depending of salinity stratal water of Romaschkino oil field

Salinity of stratal water, g/l	The number of bacteria, cell/ml				Rate of	
	SB	HOB	SRB	MB	Sulfatereduction, ηr S l/day	Methano-genesis, ηr C l/day
≤ 50	$\leq 10^3$	$\leq 10^2$	$\leq 10^4$	$\leq n\times10^2$	$\leq n\times10^4$	$\leq 10^3$
≤ 200	$\leq 10^1$	$\leq 10^1$	$\leq 10^2$	$\leq 10^1$	$\leq n\times10^3$	$\leq 10^2$
> 200	$n\times10^0$	$n\times10^0$	$n\times10^0$	$n\times10^0$	$\leq n\times10^2$	$\leq 10^1$

SB - saprotrophic bacteria
HOB- hydrocarbon-oxidizing bacteria
SRB - sulfate-reducing bacteria
MB - methanogenic bacteria

The biogeotechnology of improved oil recovery (BIOR) that we developed is based on the activation of the geochemical activity of microorganisms. Between 1983 and 1992, the BIOR was tested on a pilot scale in the Devonian horizons of the Bondyuzhskoye and Romashkino Oil Fields. These horizons were exploited for more than a year by employing the technology of flooding with freshwater, which resulted in the development of a microbial community (biocenosis) that was dominated by aerobic hydrocarbon-oxidizing microorganisms and anaerobic methanogenic bacteria.

Our attempts to activate this microbial community brought about some changes in its composition (Table 2). The introduction of oxygen and mineral nitrogen and phosphorus salts into the stratal water activated all microbiological processes in the near-bottom zone. In the aerobic region of this zone, the number of hydrocarbon-oxidizing bacteria increased by two–three orders, the content of acetate in the medium rose by an order, the content of the heavy carbon isotope ^{13}C in the dissolved carbonates somewhat decreased and that in the methane produced by methanogens somewhat increased. These changes indicated a predominance of the aerobic processes of hydrocarbon oxidation in this region. In the anaerobic region of the injection zone, the number of anaerobic sulfate-reducing and methanogenic bacteria and the rate of the terminal oxidation of oil brought about by these bacteria increased by two–three orders, the content of acetate in the medium declined by an order, and the content of the heavy carbon isotope ^{13}C in the dissolved carbonates somewhat increased and than in the methane produced by methanogens somewhat decreased. These changes indicated a predominance of the anaerobic processes of hydrocarbon oxidation in the anaerobic region. In general, the experimental data obtained in this study show a feasibility of activating microbiological processes in the near-bottom zones of injection well of oil-bearing strata into which freshwater was injected.

Table 2. The characteristic of near-bottom zone of injection well microbial community before and after activation (fresh water).

Composition of microbial community	Number of microorganisms, cell/ml			Rate of		Content of acetate, mg/l	Volume δ ^{13}C, ‰	
	HOB	SRB	MB	Sulfate-reduction, ηг S l/day	Methano-genesis, ηг C l/day		carbonates	methane
Aerobic zone of near-bottom zone of injection well								
Nature	≤ 10^3-10^4	≤ 10^1	≤ 10^1	0	0	≤1	-13÷-7	-52 ÷ -54
After activation	≤ 10^5-10^6	0	0	0	0	≤ 10-20	-20 ÷-15	-45 ÷ -51
Anaerobic zone of near-bottom zone of injection well								
Nature	≤ 10^2	≤ 10^3	≤ 10^2	≤n×10^2	≤ n×10	≤ 2-4	-11÷-4	-52 ÷ -54
After activation	≤ 10^4	≤ 10^6	≤ 10^4	≤n×10^5	≤ n×10^3	≤ 10	-15 ÷+4	-54 ÷ -65

Table 3. Change of sulfate-reducing and methanogenic bacteria number in stratal water of the Chishminskaya and Severo-Aznakaevskaya areas during 1992 – 2000 гг.

Sulfate-reducing bacteria, cell/ml			Methanogenic bacteria, cell/ml					
			from CO_2			from CH_3COO^-		
* background	** activation period	*** stabilization period	* back-ground	** activation period	*** stabilization period	* back-ground	** activation period	*** stabilization period
Chishminskaya area								
0÷10	10^2÷10^4	0÷100	0÷100	10÷10^3	0÷10	0÷10	10÷10^3	0÷1
Severo-Aznakaevskaya area								
0÷10	10÷100	0÷10	0÷100	10÷100	0÷1	0÷1	1÷10	0÷1

* - 1993.; ** - 1994÷1997.; *** - 1997÷2000.

We also investigated the downhole regions of the production wells that were located in the zone of the probable hydrodynamic influence of the injection wells. The results of these experiments, which were performed between 1992 and 2000 in the Chishminskaya and Severo-Aznakaevskaya areas, showed that a sufficiently long initial period of activation of sulfate-reducing and methanogenic bacteria in the stratal water (in this case, the number of these bacteria increased by 2–3 and 1–2 orders, respectively) was followed by a period of a gradual decrease in microbial activity to a background level.

It should be noted that the Romashkino Oil Field is primarily exploited with the injection of highly mineralized wastewater with a salinity of 120–150 g/l. The microbiological investigation of samples from the injection wells into which highly mineralized wastewater was injected for some years showed that the microbial cenosis was not formed there, the aerobic zone being entirely absent, and the anaerobic zone containing only a small number of sulfate-reducing bacteria (Table 4). Our attempts to activate the development of microbial communities by the BIOR led to an insignificant increase in the number of hydrocarbon-oxidizing and sulfate-reducing bacteria (to 10^2 and 10^3 cells/ml, respectively). To obtain a more active biocenosis, we attempted to colonize with bacteria the ground around near-bottom zone of injection well 7824 (the Tashliyar area), into which highly mineralized wastewater with a salinity of up to 150 g/l was injected for several years. For this purpose, an aerated suspension of halotolerant hydrocarbon-oxidizing microorganisms containing 10^5 cells in ml of a solution of phosphorus and nitrogen salts was injected into this well. The hydrocarbon-oxidizing microorganisms of the genera *Rhodococcus*, *Candida*, and *Pseudomonas*, which are the principle components of the commercial preparation Devoroil, were isolated earlier from the Bondyuzhskoye Oil Field.

Table 4. The characteristic of near-bottom zone of injection well microbial community (wastewater) the Tashliyar area.

Composition of microbial community	The number of bacteria, cell/ml		
	HOB	SRB	MB
Aerobic zone of near-bottom zone of injection well			
Nature	$n \times 10^0$	0	0
After treatment (well 7824)	10^4	0	0
Anaerobic zone of near-bottom zone of injection well			
Nature	$n \times 10^0$	$\leq 10^2$	0
After treatment (well 7824)	10^2	$\leq 10^3$	≤ 10

Measurements performed one year thereafter showed the presence of hydrocarbon-oxidizing microorganisms in the aerobic region of the injection zone inoculated with the microorganisms in an amount of 10^4 cells/ml and the occurrence of the terminal processes of the microbiological oxidation of oil (sulfate reduction and methanogenesis). In other words, these experiments showed a feasibility of creating appropriate microbial communities in the zone of injection of highly mineralized wastewater.

Similar experiments were performed on the Bobrikovsky horizon (well 26142, oil deposit 31), whose stratal water is highly mineralized and contains hydrogen sulfide. The oil of this horizon is more viscous than the Devonian oil. The results of the comprehensive microbiological studies of this horizon are presented in Tables 5 and 6, which also contain, for the sake of comparison, data for the Devonian horizon in the same area.

Table 5. The number of aerobic bacteria in stratal water of the Bobrikovsky horizon and the Devonian horizon (Chishminskaya area) (cell/ml).

Horizon	Bacteria, oxidizing		Saprotrophes, isolated from	
	Crude oil	n-alkanes	Water	Crude oil
Bobrikovsky	0 - 1	0 - 100	0 - 50	0 - 160
Devonian	0 - 100	0 - 100	0 - 20	0 - 100000

Table 6. The number of anaerobic bacteria in stratal water of the Bobrikovsky horizon and the Devonian horizon (Chishminskaya area) (cell/ml).

Horizon	Methanogenes			Acetogenic bacteria	SRB
	$H_2 : CO_2$	CH_3COO^-	MTMA		
Bobrikovsky	0 - 1	0 –1	0 - 1	0 - 10	0 - 10
Devonian	0 – 10	0 –1	0 -100	1 -10	0 -100

As can be seen from these tables, the stratal water of the Bobrikovsky horizon contained aerobic microorganisms, although their number was 1–2 orders less than in the stratal water of the Devonian horizon. The number of hydrocarbon-oxidizing microorganisms typically did not exceed several cells in ml. Heterotrophic microorganisms were detected not only in the stratal water but also in the oil of the Bobrikovsky horizon, their number being greater than the number of hydrocarbon-oxidizing microorganisms. The presence of aerobic microorganisms in the anaerobic environment can be explained solely by their inoculation with the injected water. The data obtained indicate that there is a good hydrodynamic contact between the production and injection wells in the Bobrikovsky horizon. It should be noted that aerobic microorganisms (including hydrocarbon-oxidizing bacteria) in the injection zones were suppressed, which is likely to be related to a deficiency of oxygen in this environment.

The number of anaerobic microorganisms in the stratal water of the Bobrikovsky horizon, as a rule, was 1–2 orders less than in the stratal water of the Devonian horizon. The number of methanogenic bacteria in samples from the former horizon was as low as several cells in ml (if at all), whereas the number of acetogenic and sulfate-reducing bacteria in some samples reached 10 cells/ml. Analysis showed that the injected wastewater contained no methanogens and only a few cells of acetogenic and sulfate-reducing bacteria in ml. The analysis of the stratal water issuing from different depths of the injection wells showed that the number of anaerobic microorganisms virtually did not increase with the depth.

In contrast to the Devonian horizon, the creation of active microbial communities in the Bobrikovsky horizon by the BIOR required, because of the presence of hydrogen sulfide in the latter horizon, the introduction of an oxidizer with the injected water. For this reason, before applying the basic BIOR, the hydrogen sulfide present in the Bobrikovsky horizon was neutralized by adding a sufficient amount of an oxidizer to the injected water. The analysis of the stratal water issuing from the injection well several months after the injection of the oxidizer showed the presence of live cells of hydrocarbon-oxidizing bacteria in the injection zone.

The experimental data obtained in this study show that the stratal water of the Bobrikovsky horizon is not sterile, at least at the current exploitation stage of oil deposit 31. The anaerobic component of the trophic chain of oil oxidation in this horizon is functional, whereas the aerobic component is suppressed. There are grounds to believe that not only anaerobic but also aerobic microbiological processes in this horizon can be activated to levels typical of the Devonian horizon.

References

1. Ivanov, M.V., Belyaev, S.S., Laurinavichus, K.K., Obraztsova, A. Ya., and Gorlatov, S.N.: The Distribution and Geochemical Activity of Microorganisms in a Waterflooded Oil Deposit, *Mikrobiologiya*, 1982, vol. 51, no. 2, pp. 336–341.
2. Milekhina, E.I., Borzenkov, I.A., Belyaev, S.S., Miller, YU.M., and Ivanov, M.V.: The Hydrocarbon-oxidizing Microflora of Waterflooded Oil Deposits in Tatarstan with Different Mineralization of Stratal waters, *Mikrobiologiya*, 1991, vol. 60, no. 4, pp. 747–755.
3. Belyaev, S.S., Borzenkov, I.A., Milekhina, E.I., Charakhchyan, I.A., AND Ivanov, M.V.: The Dynamics of Microbiological Processes in the Exploited Horizons of the Romashkino Oil Field, *Mikrobiologiya*, 1990, vol. 59, no. 6, pp. 1118–1126.
4. Raymond R.L.: Microbial oxidation of n-parafinic hydrocarbons // *Develop. Industr. Microbiol.* 1961. №2. P. 23-29.
5. Posgate J.R.: Media for sulfur bacteria // *Laboratory practice*. 1966. N15. P.1239-1244.
6. Trüper H.G., Schlegel H.G.: Sulphur metabolism in Thiorhodaceae. I. Quantitative measurements of growing cell of Chromatium okenii. // Antonie van Leeuwenhoek., *J. Microbiol. Serol.* 1964. V. 30. №3. P. 225-238.
7. Oremland R.S., Capone O.G.: Use of specific inhibition in biogeomicrobial ecology // *Adv. Microbiol. Ecol.* 1988. V. 10. P. 285-384.
8. Kenely W., Zeikus J.G.: One-carbon metabolisms in methanogen: evidence for synthesis of a two-carbon cellular intermediate and unification of catabolism in Methanosarcina barkery // *J. Bacteriol.* 1981. V. 146. P. 133-140.
9. Belyaev S.S.,Wolkin R., Kenlay W.R., Deniro M.J., Epstein S., Zeikus J.G.: Methanogenic bacteria from the Bondyuzhskoe oil field: general characterization and analysis of stable-carbon isotopic fractionation // *Appl. Environ. Microbiol.* 1983. V. 45. P. 1239-1244.
10. Charakhchyan, I.A., Mityushina, L.L., Kuznetsova, V.G., and Belyaev, S.S.: The Morphology and Ultrastructure of Methanococcoides euhalobius an the Role of Calcium in Its Morphogenesis, *Mikrobiologiya*, 1989, vol. 58, no. 5, pp. 825–830.
11. Borzenkov, I.A., Belyaev, S.S., Miller, Yu.M., Davydova, I.A., and Ivanov, M.V.: Methanogenesis in the Highly Mineralized Stratal Waters of the Bondyuzhskoye Oil Field, *Mikrobiologiya*, 1997, vol. 66, no. 1, pp. 122–129.
12. Laurinavichus, K.K. and Belyaev, S.S.: The Radioisotope Assay of the Microbiological Production of Methane, *Mikrobiologiya*, 1978, vol. 47, no. 6, pp. 1115–1117.

13. Esikov, A.D.: Methodological Recommendations on the Precise Mass Spectrometric Determination of the Isotope Composition of Carbon and Hydrogen, Moscow: *Inst. Nucl. Geophys. Geochem.*, 1975.

INDEX

O

P

R

S

T

V

W

X

Y

Z